Advanced Bioinspiration Methods for Healthcare Standards, Policies, and Reform

Hadj Ahmed Bouarara
GeCoDe Laboratory, Dr. Moulay Tahar University, Algeria

A volume in the Advances in Healthcare Information Systems and Administration (AHISA) Book Series

IGI Global
PUBLISHER of TIMELY KNOWLEDGE

Published in the United States of America by
IGI Global
Medical Information Science Reference (an imprint of IGI Global)
701 E. Chocolate Avenue
Hershey PA, USA 17033
Tel: 717-533-8845
Fax: 717-533-8661
E-mail: cust@igi-global.com
Web site: http://www.igi-global.com

Library of Congress Cataloging-in-Publication Data

Names: Bouarara, Hadj Ahmed, 1990- editor.
Title: Advanced bioinspiration methods for healthcare standards, policies, and reform / Hadj Ahmed Bouarara, editor.
Description: Hershey PA : Medical Information Science Reference, [2023] | Includes bibliographical references and index. | Summary: "This book is the fusion of AI and Medicine bringing together the latest empirical research findings in the areas of AI, BI, law, ethics and medicine to help readers optimize the potential benefits of Artificial Intelligence models and BioInspired algorithms in health issues while mitigating potential dangers by examining the complex issues, and innovative solutions that are linked to healthcare standards, policies, and reform"-- Provided by publisher.
Identifiers: LCCN 2022023237 (print) | LCCN 2022023238 (ebook) | ISBN 9781668456569 (hardcover) | ISBN 9781668456576 (ebook)
Subjects: MESH: Artificial Intelligence | Algorithms | Artificial Intelligence--ethics | Confidentiality
Classification: LCC R859.7.A78 (print) | LCC R859.7.A78 (ebook) | NLM W 26.55.A7 | DDC 362.10285--dc23/eng/20220727
LC record available at https://lccn.loc.gov/2022023237
LC ebook record available at https://lccn.loc.gov/2022023238

This book is published in the IGI Global book series Advances in Healthcare Information Systems and Administration (AHISA) (ISSN: 2328-1243; eISSN: 2328-126X)

British Cataloguing in Publication Data
A Cataloguing in Publication record for this book is available from the British Library.

All work contributed to this book is new, previously-unpublished material.
The views expressed in this book are those of the authors, but not necessarily of the publisher.

For electronic access to this publication, please contact: eresources@igi-global.com.

Advances in Healthcare Information Systems and Administration (AHISA) Book Series

ISSN:2328-1243
EISSN:2328-126X

Editor-in-Chief: Anastasius Moumtzoglou Hellenic Society for Quality & Safety in Healthcare and P. & A. Kyriakou Children's Hospital, Greece

MISSION

The **Advances in Healthcare Information Systems and Administration (AHISA) Book Series** aims to provide a channel for international researchers to progress the field of study on technology and its implications on healthcare and health information systems. With the growing focus on healthcare and the importance of enhancing this industry to tend to the expanding population, the book series seeks to accelerate the awareness of technological advancements of health information systems and expand awareness and implementation.

Driven by advancing technologies and their clinical applications, the emerging field of health information systems and informatics is still searching for coherent directing frameworks to advance health care and clinical practices and research. Conducting research in these areas is both promising and challenging due to a host of factors, including rapidly evolving technologies and their application complexity. At the same time, organizational issues, including technology adoption, diffusion and acceptance as well as cost benefits and cost effectiveness of advancing health information systems and informatics applications as innovative forms of investment in healthcare are gaining attention as well. **AHISA** addresses these concepts and critical issues.

COVERAGE

- IT Applications in Health Organizations and Practices
- IT security and privacy issues
- Measurements and Impact of HISA on Public and Social Policy
- IT Applications in Physical Therapeutic Treatments
- Role of informatics specialists
- Virtual health technologies
- Clinical Decision Support Design, Development and Implementation
- Rehabilitative Technologies
- IS in Healthcare
- E-Health and M-Health

IGI Global is currently accepting manuscripts for publication within this series. To submit a proposal for a volume in this series, please contact our Acquisition Editors at Acquisitions@igi-global.com or visit: http://www.igi-global.com/publish/.

The Advances in Healthcare Information Systems and Administration (AHISA) Book Series (ISSN 2328-1243) is published by IGI Global, 701 E. Chocolate Avenue, Hershey, PA 17033-1240, USA, www.igi-global.com. This series is composed of titles available for purchase individually; each title is edited to be contextually exclusive from any other title within the series. For pricing and ordering information please visit http://www.igi-global.com/book-series/advances-healthcare-information-systems-administration/37156. Postmaster: Send all address changes to above address.

Titles in this Series

For a list of additional titles in this series, please visit:
www.igi-global.com/book-series/advances-healthcare-information-systems-administration/37156

Health Informatics and Patient Safety in Times of Crisis
Narasimha Rao Vajjhala (University of New York Tirana, Albania) and Philip Eappen (University of Toronto, Canada)
Medical Information Science Reference • © 2023 • 286pp • H/C (ISBN: 9781668454992) • US $345.00

Digital Identity in the New Era of Personalized Medicine
Ingrid Vasiliu-Feltes (University of Miami, USA) and Indira Mysore (Bloqcube Inc., USA)
Medical Information Science Reference • © 2023 • 315pp • H/C (ISBN: 9781799889663) • US $325.00

Assessing the Need for a Comprehensive National Health System in the United States
Nikolaos Karagiannis (Winston-Salem State University, USA) Sheilia R. Goodwin (Winston-Salem State University, USA) and David Stewart (Winston-Salem State University, USA)
Information Science Reference • © 2023 • 300pp • H/C (ISBN: 9781668440605) • US $250.00

Cases on Lean Thinking Applications in Unconventional Systems
Eduardo Guilherme Satolo (São Paulo State University, Brazil) and Robisom Damasceno Calado (Fluminense Federal University, Brazil)
Engineering Science Reference • © 2023 • 262pp • H/C (ISBN: 9781668451854) • US $240.00

Handbook of Research on Complexities, Management, and Governance in Healthcare
Ubaldo Comite (University "Giustino Fortunato" Benevento, Italy)
Medical Information Science Reference • © 2023 • 411pp • H/C (ISBN: 9781668460443) • US $470.00

For an entire list of titles in this series, please visit:
www.igi-global.com/book-series/advances-healthcare-information-systems-administration/37156

IGI Global
PUBLISHER of TIMELY KNOWLEDGE

701 East Chocolate Avenue, Hershey, PA 17033, USA
Tel: 717-533-8845 x100 • Fax: 717-533-8661
E-Mail: cust@igi-global.com • www.igi-global.com

Editorial Advisory Board

Table of Contents

Detailed Table of Contents

In December 2019, an outbreak of severe acute respiratory syndrome coronavirus 2 (SARSCoV-2) infection occurred in Wuhan, Hubei Province, China and spread across China and beyond. On February 12, 2020, WHO officially named the disease caused by the novel coronavirus as Coronavirus Disease 2019. In the literature, COVID-19 is characterized by the symptoms of viral pneumonia such as fever, fatigue, dry cough, and lymphopenia. Many of the older patients who become severely ill have evidence of underlying illness such as cardiovascular disease, liver disease, kidney disease, or malignant tumors. The COVID-19 pandemic posed a great danger at all levels, and despite the doctors' rush to find a vaccine, it will remain a threat to the world unless people adhere to health precaution. The discussion in this chapter of health precautions will focus largely on two main components: wearing face masks and respecting social distancing, which is the distance regulated by the World Health Organization (1 meter); also it will discusses the symptoms caused by this pandemic and its impact on some areas.

The severe acute respiratory syndrome coronavirus 2 (SAR-Cov-2) caused the coronavirus (COVID-19) pandemic. The global concern is the discovery of a new target drug for the total cure. Recently, some research showed that a few COVID-19 enzymes may have been contemplated to be potential drug targets, but not much is known about its structural biology. This research investigates the 3-D structure of protease SAR-CoV-2. The tertiary structure was determined by homology modeling. The Swiss-Model workspace and the basic local alignment search tool (BLAST) were employed for modeling, and the resulted model was validated with programs that include PROCHECK, Verify3D, and QMEAN to ascertain reliability. To establish the structures that fitted, HHBlits software was employed. To verify the structure quality, a Ramachandran plot was exploited. The binding site was determined using CASTp and DeepSite algorithms, which showed that this protein can be exploited as a prospective pharmaceutical target. Albeit further experimentation is required as a COVID-19 virus vaccine-design/target-drug.

Informatics and internet technologies are becoming extremely popular in today's healthcare system. The emergence of the worldwide web has affected the way in which health-related information is distributed and accessed over cyberspace. The internet is rapidly gaining importance, not just for healthcare professionals, but also for patients, by enabling them to search for drug- and other health-related information. Bioinformatics is the combination of biology and information technology. The term bioinformatics was coined by Paulien Hogeweg in 1979 for the study of informatic processes in biotic systems. Its primary use since the late 1980s has been in genomics and genetics, particularly in those areas of genomics involving large-scale DNA sequencing.

Computer vision has made substantial progress during the past several decades, and it has covered many areas including healthcare. This chapter introduces deep learning techniques for healthcare with an emphasis on the discussion of deep learning in computer vision, the authors describe how these techniques can impact a few key areas of health precaution that include wearing a mask and respecting social distancing to reduce the spread of airborne epidemics. This chapter provides an overview of deep learning techniques that can detect social distancing and mask wearing through modern surveillance technologies, including drones and surveillance cameras.

A number of fundamental rights are inseparable from human personality. These are in particular the right to the name, the right to missive letters, the right to physical integrity, the moral right of the author and, in the field which interests us, the right to the image, the right to honor, or the "right to secrecy". These rights even have constitutional value. The Constitutional Council considers that they proceed from "respect for individual freedom." In particular, laws that violate the search of vehicles, tax searches, and the publication of certain elements of tax declarations, have been declared unconstitutional. In this chapter the authors will talk about the different techniques that can be used for the preservation of privacy. This chapter will illustrate an overview of the different fundamental techniques for the preservation of privacy in medical sector such as: differential privacy, secure multiparty computation protocol, data anonymization, deidentification, fingerprint privacy, and others.

Large-scale research evaluations, as well as the acquisition and analysis of massive amounts of data, in order to follow trends and models form the organization. As the world around us continues to change, many entrepreneurs are interested in learning how big data might help them enhance their businesses. While businesses have traditionally analyzed data, recent technical advancements have spawned new entrants and unlocked the potential of big data. The extent of the consequences of big data, according to an SNS research, is presently $57 billion and expected to rise further.

Healthcare, which is considered as one of the most strategic fields, has noticed a remarkable and continuous progress over time. Yet, this sensitive field though the crucial role it plays in providing both individuals and communities with cure and care, still needs more attention from policy makers, researchers, and doctors due to the rapid and intensive progress information communication technologies (henceforth ICT) improve day after day. Hence, the present investigation attempts to highlight the use of ICT technologies in developing healthcare in Algeria as a case study and seeks to suggest better ways of using technological advancement to overcome the challenges that still hinder against its progress. The study in hand is divided into several parts. First, it tackles the recent situation of healthcare in matter of using digital technologies in accessing, processing, and storing information. Then, it tries to shed light on the different drawbacks of the misuse of ICT in healthcare settings and state the causes and consequences.

Foreword

The concept of bio-inspiration appears more than ever as a major trend in artificial intelligence for innovation. The idea is to observe and better understand nature to replicate its genius. Healthcare is a field much closer to life and biology than many others. The link between health and biomimicry is therefore quite natural. However, it is clear that biomimicry is effective in improving many aspects of health.

The Bio-inspiration is also a branch of engineering sciences that consists of drawing inspiration from nature to improve or develop respectful medical products while being very efficient. Applications of bio-inspired engineering are found today in a large number of industrial fields: aeronautics, materials, IT, robotics, artificial intelligence, energy, housing, etc. However, they would currently represent only a little more than 10% of the technologies developed by man. Several scientists, commissions and task forces predict that artificial intelligence (AI) will transform and democratize healthcare systems by improving the quality, safety and efficiency of care, and offer patients the tremendous opportunity to take control of their own health. Biomimicry can improve human health by enhancing systems and service delivery. Well-designed algorithms could, for example, correct the currently high rates of preventable medical errors. Yet, at the same time, the use of AI and bioinspiration in healthcare presents obvious risks and implementation issues, such as those related to discrimination, informed consent, safety or quality. (and liability for damages caused) as well as the protection of privacy. The absence of regulations to mitigate these risks could compromise our society's ability to realize the potential benefits of bioinspiration in healthcare. For example, AI and bioinspiration are unlikely to reduce medical errors if it has been trained on unrepresentative (and therefore biased) data that causes such errors. The use of low-quality AI or bioinspiration like neural networks in health is likely to lead to backlash from the public and burdensome regulations that would stifle innovation. The combination of two major fields bioinspiration and health care represents the fusion of law and science.

The presented book through its different chapters Answers a general question which is how can we optimize the potential benefits of bioinspiration (like deep learning techniques) in the service of health while mitigating the potential dangers?

It aims to enrich literature by new original scientific contribution that focus on the challenges presented by privacy, security and quality issues in regulations and on ethical issues related to the use of health AI. Going through chapters, reader will enjoy exploring a collection of different bioinspired and AI applications on different subdomains of Healthcare.

Benyahia Kadda
Dr. Moulay Tahar University, Algeria

Preface

The use of bioinspiration technologies is effective in the evaluation of pathologies to be treated. It helps physicians in the analysis and interpretation of patient data. Today, bioinspiration and AI in general are becoming the doctor's ally. They make it possible to remotely measure body constants such as heart rate, saturation, respiratory rate, etc. The camera makes it possible to identify pallor, cyanosis, weight loss, changes in the texture of the skin imperceptible to the naked eye. Bioinspired techniques optimize the detection of symptoms and promotes predictive monitoring of the deployment of a disease. This ambitious technology promises help in diagnosis, decision-making, personalization of treatments and self-management of pathologies.

The bioinspiration in medicine has infinite potential. Its limits are constantly pushed back and the fields of application are numerous. Health actors hope that the solutions provided by AI and bioinspiration will limit medical errors. The health sector must therefore respond to strategic challenges in terms of training health professionals. Within hospitals, all staff must be able to understand the operation, benefits, and limits of bioinspiration. Artificial intelligence is also present in the self-management of patient health. Thanks to AI, the patient is now more involved in the monitoring and management of his disease.

BOOK ORGANISATION

The book content is organized as follow:

The first chapter, "The Ethical Consideration of Using Artificial Intelligence (AI) in Medicine," aims to discuss the ethical consideration of using AI in medicine. Some medical decisions must involve ethical principles, such as whether to withhold treatment from a patient as well as the infusion of AI and medicine.

In the second chapter, "Advanced Bioinspiration Methods for Optimisation Problems: Survey and Applications," the authors talk about metaheuristics and bio-inspired algorithms in general. From their basic concepts to their applications.

The content of the third chapter, “Health Precautions for Patient Safety,” elaborates the discussion of health precautions will focus largely on two main components: wearing face masks and respecting social distancing, which is the distance regulated by the World Health Organization (1 meter), also discusses the symptoms caused by this pandemic and its impact on some areas.

The principal objective of the fourth chapter, “Homology Modeling and Binding Site Analysis of SARS-CoV-2 (COVID-19) Main Protease 3D Structure,” is to investigate the 3-D structure of protease SAR-CoV-2. The tertiary structure was determined by homology modeling. The Swiss-Model workspace and the basic local alignment search tool (BLAST) were employed for modeling, and the resulted model was validated with programs that include PROCHECK, Verify3D, and QMEAN to ascertain reliability.

The fifth chapter, “Convolutional Neural Networks (CNN): Models and Algorithms,” will encompass the different models of the convolutional neural network that exist in the literature, their parameters, their architectures, their layers, their advantages, and disadvantages. The objective is to help researchers to have a summary on deep learning.

The sixth chapter, “A Hybrid Classification Algorithm and Its Application on Four Real-World Data Sets,” which proposes a hybrid classification algorithm based on particle swarm optimization (PSO) to enhance the generalization performance of the adaptive boosting (AdaBoost) algorithm, is very interesting.

In the seventh chapter, “Advanced Machine Learning Techniques in Spam Medical Events: Review of Literature,” the primary objective is to discuss the different algorithms of machine learning such as the KNN Decision tree, naive bayes, etc. applicable to the filtering of medical spam event. The objective of these techniques is to control medical data in order to make decisions and achieve strategic objectives.

The eighth chapter, “Artificial Intelligence in the Healthcare Sect,” opens doors for discussing the trending areas of technical capability of AI in the healthcare sector in general and its areas of specialization in specific, giving a specific example of solving a prediction problem at the end of the chapter.

The purpose of the ninth chapter, “Genetic Algorithm and Machine Learning,” was to combine genetic algorithms with machine learning techniques.

The authors in the tenth chapter, “Application of Bioinformatic in Cancer Prediction and Prognosis,” show an overview of the different bioinformatics techniques such as image segmentation as well as the most used and popular medical datasets existed in literature

The eleventh chapter, “Computer Vision in Healthcare,” provides an overview of deep learning techniques that can detect social distancing and mask wearing through modern surveillance technologies, including drones, surveillance cameras, etc.

In the twelfth chapter, "Privacy Preserving in the Medical Sector: Techniques and Applications," the authors illustrate an overview of the different fundamental techniques for the preservation of privacy in medicale sector such as differential privacy, secure multiparty computation protocol, data anonymization, deidentification, fingerprint privacy, etc.

The thirteenth chapter, "Technological Innovation Driven by Big Data," discusses how big data might help medical entrepreneurs to enhance their businesses.

The fourteenth chapter, "Towards an Effective Integration of ESP in Healthcare ICTs: Challenges and Strategies – The Case of Algeria," is divided into three parts. First, it tackles the recent situation of healthcare in matter of using digital technologies in accessing, processing, and storing information. Then, it tries to shed light on the different drawbacks of the misuse of ICT in healthcare settings and state the causes and consequences. Finally, the study tends to suggest fruitful strategies so as to improve a good use of new technologies within the domain of healthcare particularly in Algeria.

Hadj Ahmed Bouarara
GeCoDe Laboratory, Dr. Moulay Tahar University, Algeria

Acknowledgment

I have taken effort in this project. However it would not have been possible without the support of many individuals. I would like to extend my sincere thanks to all of them.

First and foremost, I would like to thank the reviewers for their help comments and suggestions as well as the teachers of the Dr. Molay Tahar Saida University, Algeria, and my colleagues of the laboratory GeCoDe who helped me in order to finalize this project.

I would like to express my gratitude to the global manager team of IGI global for their suggestions and helps in order to make our work much less painful than it could be.

Thank you finally to the authors who will recognize themselves here, for their participation so precious and especially for their availability which allowed me to realize in good conditions this project.

Chapter 1
The Ethical Consideration of Using Artificial Intelligence (AI) in Medicine

Rita Komalasari
https://orcid.org/0000-0001-9963-2363
Yarsi University, Indonesia

ABSTRACT

In the last decade, intelligent systems are becoming more popular. In this chapter, the authors will examine the ethical consideration of using intelligent systems for medicine. artificial intelligence (AI) can help health workers provide a better level of care for their patients. This paper aims to discuss the ethical consideration of using AI in medicine. Some medical decisions must involve ethical principles, such as whether to withhold treatment from a patient as well as the infusion of AI and medicine.

INTRODUCTION

This chapter contains reviews based on the concept of the book *Advanced Bioinspiration Methods for Healthcare Standards, Policies, and Reform*. Both academics and the general public will benefit significantly from this chapter—for example, a government official dealing with intelligent systems for medicine.

In the last decade, intelligent systems have become more popular (Dewangan, 2019). Four ethical issues—accuracy, accessibility, privacy, and ownership—are the focus of this chapter. In this section, we will take a closer look at the ethical consideration of using intelligent systems for medicine. The first section discusses

DOI: 10.4018/978-1-6684-5656-9.ch001

technical arguments about whether intelligent systems should be used in medical treatment. Since this is a quickly developing field, we picked current papers as our methodology. Half of it has been released in the prior five years. We take some consideration for issues that arise when building intelligent systems. The second section provides a range of ethical arguments about whether health professionals should use intelligent systems in medicine. This section summarises four ethical considerations. Accuracy, accessibility, privacy, and ownership. In the third part of this chapter, we discuss the papers we found while looking for existing taxonomies that other researchers of intelligent medical systems have already made.

METHOD

Methodologically, we picked sensitive search words (finding the most crucial studies but with much irrelevant material) after the initial searches (more relevant research but with the risk of missing some important ones). During the article's analysis, we used the phrase "intelligent systems" in this section. The comprehensive, evidence-based examination covers intelligent medical systems. We chose recent publications since this is a rapidly expanding area. They have published it half in the previous five years. After evaluating abstracts, we estimated 35 texts.

We also searched on the Google search engine for 'Artificial Intelligence routine use.' This search returned another website (www.openclinical.org/aisinpractice.html) as being relevant. This site, which had not existed the previous year, held a similar list of artificial intelligence systems. An acknowledgment on this page said that Professor Coiera had initially compiled the archive; we used it with his permission; and that it was now maintained and updated by OpenClinical. The list was longer than the original website; OpenClinical had continued searching for systems. Professor Coiera's website listed the status of some 39 applications, while the OpenClinical website had increased this number to 50. Importantly, these numbers do not merely reflect the current systems in clinical practice. Professor Coiera's website also showed information on systems that were of "unknown status," "under evaluation," "decommissioned," or "prototype."

The OpenClinical website also listed all these systems except 'prototype'; the one 'prototype' entry from the previous list, ADE Monitor, is now listed as unknown. Of the 39 applications listed on Professor Coiera's website, we listed 25 as "routine use." Only 12 of the 50 systems listed on the OpenClinical website were in "routine use." Eight of the 12 systems listed as routinely used were exclusive to the OpenClinical list; we included only four of the systems listed as routinely used from Professor Coiera's list of 25. Two systems shown as being in routine use by Professor Coiera's earlier list were listed as being "decommissioned." OpenClinical still listed the remaining 19 systems listed by Professor Coiera as being in regular service, but they had no status description. We assume that OpenClinical knew that these 19 systems were in routine use but can no longer be sure if they still are.

Of the 39 applications listed on Professor Coiera's website, we listed 25 as "routine use." Only 12 of the 50 systems listed on the OpenClinical website were in "routine use." Eight of the 12 systems listed as routinely used were exclusive to the OpenClinical list; we included only four of the systems listed as routinely used from Professor Coiera's list of 25. Two systems shown as being in routine use by Professor Coiera's earlier list were listed as being "decommissioned."

The literature search resulted in 25 systems being in routine clinical use. OpenClinical, which had worked more recently on this problem, lists 12 methods as definitely being in everyday clinical use, with another three systems being of similar status (e.g., "in clinical use"). This search confirmed the existence of 24 designs in routine clinical use. Unfortunately, it is impossible to find out how many of the systems listed by OpenClinical were the same ones as those in this search. The names of all the systems detected in this search are not known. We only described many methods in this search regarding how they work and what they can do.

The scarcity of applications in this search suggested that we had missed many applications, or there were very few of them in routine use. These figures suggest that this search probably found a significant proportion of the intelligent systems in regular clinical use worldwide and that there are not very many of them. The results presented in this chapter use quality assurance by enhancing the credibility and veracity of the data. We made specific facts that supported each subject and explanation.

RESULTS AND DISCUSSION

The following section presents a result and discussion of the ethical consideration of using intelligent systems for medicine. Many researchers have investigated whether we use ethical computer systems in many fields (Maedche et al., 2019). Still, few have focused on the specific case of using intelligent systems in medicine (Kaur & Chana, 2014). Intelligent systems have unique issues associated with them. They mimic the intelligent behavior of humans. This mimic may offend those whose intelligence mimics; some may believe that developing systems capable of replacing human intellectual work can be morally wrong (Sparrow, 2007).

Computer ethicists have argued about computer systems' use or non-use. This section looks at those that may have something to do with whether we should use intelligent systems in medicine.

Table 1. Ethical considerations of intelligence system

Privacy		Accuracy		Ownership		Access	
pros	cons	pros	cons	pros	cons	pros	cons
the dissemination increases the necessity of safeguarding a patient's right to privacy	Electronic patient data recording may jeopardise a person's fundamental right to privacy	using bedside computer terminals to chart care plans and medication details increased the accuracy of the data recorded	the interconnectedness of networked and computerized record keeping will increase the chances of misuse or error	The protection of property rights encourages creativity and invention, which makes available new products and processes	The software license may violate "the notion of mental stages,"	This program would give all citizens the opportunity, ability, and desire to use electronic resources	equality between the "information-rich" and the "information poor."

This section will start by reviewing four issues identified by Mason (2017). 'PAPA' is a concept coined by Mason in their book 'Four Ethical Issues of the Information Age.' Accuracy, accessibility, privacy, and ownership are PAPA's core values, and intelligent systems' use may affect these ethical issues (Young et al., 2020). A Table. 1 explaining some of the pros and cons for each ethical consideration.

Privacy

Privacy has been critical to physicians for a long time (Keshta & Odeh, 2021). The Hippocratic Oath has been binding on doctors since the 4th century BC (Askitopoulou & Vgontzas, 2018). One of the essential parts of this oath is that doctors must keep patient information secret (Veatch, 2012).

Electronic patient data recording may jeopardize a person's fundamental right to privacy (Nissenbaum, 2020). The medical record may include private information that the patient may not want to be publicized (Kaelber et al., 2008). A patient's ability to gain or keep a job or get health insurance may be affected if this information is made public, so we should keep it private (Ventola, 2014).

If we integrate patient information into a single extensive database and other data, we can infer relationships between those data and others without the patient's knowledge or permission (Ekblaw et al., 2016). The development of computer networks, including the Internet, has allowed and promoted the dissemination of private health data (Glasgow et al., 2012). According to Yaacoub et al. (2020), the dissemination increases the necessity of safeguarding a patient's right to privacy. A previous study highlights people's possible issues with protecting their private lives and human dignity while not limiting creativity [within computing] and the growth of information technology (Morley et al., 2020).

Krittanawong et al. (2020) identify a potential problem with advances in new technologies, such as genetics. It is now possible to establish traits and diseases that are hereditary. The knowledge that an individual may be at increased risk of such characteristics will affect that individual's self-perception and self-esteem. Krittanawong et al. indicate that an individual must also have the right not to know in such situations.

It is not only in the interests of the individual that we keep medical data secret (Brown, 2018). This secrecy is central to the healthcare systems' being perceived as trustworthy by the public, subject to exceptional circumstances where public health is at risk (Toreini et al., 2020). This author considers that this is a fundamental ethical principle.

Accuracy

Accuracy is a concern when there is the danger that information held electronically might be in error (Mason, 2017). It is straightforward to disseminate data across many networks to various computers quickly. If we find an error in one database, it may be difficult or impossible to rectify every database that holds erroneous information (Dilsizian & Siegel, 2014).

Coventry & Branley (2018) propounds that the interconnectedness of networked and computerized record keeping will increase the chances of misuse or error. This interconnectedness is because the volume of data being stored and moved around increases the likelihood of incorrect data being input and then propagated to other systems incorrectly. It might complicate any errors to detect and correct.

We may use an intelligent system to make potentially life-threatening decisions about a patient's treatment. The data used to make these decisions must be accurate. Li & Chapman (2020) indicate that it is impossible to reduce medical decisions to deductions based only on facts. Some medical decisions must involve ethical principles, such as whether to withhold treatment from a baby born with significant congenital disabilities. According to their findings, it would be tough for a medical expert system to include these ethical considerations.

Intelligent systems can be considered less adaptable than humans concerning the data they can use. Longoni et al. (2019) argue that information gathered from a patient by a physician will not be aimed but idiosyncratic. Individual professionals having different levels of knowledge, training, expertise, and habits cause this distinctiveness. This quirk does not work well with a system that needs a certain level of consistency to work right.

Osoba & Welser (2017) argued that their users often regard computerized records as infallible. This infallibility could cause users to reduce their vigilance when dealing with the data and may mean that they made inappropriate decisions. In contrast to these findings, Alazzam et al. (2022) reported on nurses' use of the HELP intelligent system. They found that using bedside computer terminals to chart care plans and medication details increased the accuracy of the data recorded. This system was because the nurses could enter the information immediately before they had any chance to forget details. The authors also found that the bedside terminals resulted in the more productive use of nurses' time. Cai et al. (2019) add that using computer systems can allow for better decision-making, help doctors deal with the increasing complexity of their work, and improve efficiency and consistency in the workplace (see also Musen et al., 2021).

Property

Mason's third issue concerns the property, including those intellectual property rights. To create an intelligent system, it is necessary to gather information from specialists in the industry and incorporate it into computer programs. They can then trade this software as a commodity, although they may not compensate the knowledge contributors for their input (Mason, 2017). This commodity means physicians could aid developers with the creation of intelligent systems but not receive total financial compensation for their intellectual information. The protection of property rights encourages creativity and invention, which makes available new products and processes (Cockburn et al., 2018).

A further issue related to property is discussed in Cockburn et al. (2018), where the authors discuss natural rights arguments for and against software ownership. The software license may violate "the notion of mental stages," according to a fight against the right to "own." We cannot possess a sequence of mental processes like adding or subtracting. A computer can do these operations much faster than any human, but this is essentially all a computer program does. This ability means that anything that a computer program can do, a human could also do (albeit at a much slower rate). The argument says that the ownership of computer programs may thus interfere with freedom of thought. If someone owns a computer program, they may prevent others from making the same mental steps in their heads. This argument is now discredited in the discussion of whether software can be owned. However, we can extend this argument against using artificial intelligence in medicine. One could say that physicians have the right to the freedom of thought required to treat their patients. If we forced those physicians to use artificial intelligence in their work to treat those patients, this would interfere with their freedom of thought and action.

Access

Access is the final issue highlighted by Mason. One concern is equality between the "information-rich" and the "information poor." Intelligent systems are available to those organizations with the most significant resources, primarily used by higher social classes. This usage could create a gap between the quality of healthcare provided to the rich and poor (Mason, 2017). Further, Luengo et al. (2020) identify "information literacy" as a critical requirement for users of systems. The authors argue that there is a need for the government to provide programs for education and awareness. This program would give all citizens the opportunity, ability, and desire to use electronic resources.

Brewer et al. (2005) cite developing countries as information-poor that need knowledge-based systems (KBSs) to improve their healthcare provision. They argue that KBSs could achieve greater efficiency than that which already exists; provide and maintain services of better quality than those which already live; assist in education; provide expertise when it is not possible to give an expert, and aid health workers in the identification of equipment faults and with any repairs to that equipment. Any of these problems can exist in a developing country, so that KBSs may be mainly usefully deployed there.

Further Ethical Issues

Another ethical issue affected by computers, separate from the four identified by Mason, is that of autonomy. Computer use can affect people's independence in the workplace. In their respected text 'Computer Ethics' (Floridi, 2010), Floridi puts forward the scenario of doctors using expert systems to make medical diagnoses. An expert system can be created by drawing together medical knowledge from many of the world's best doctors. If any 'ordinary' doctor were to use this system, we would significantly enhance that doctor's decision-making ability. However, this would be at the cost of reducing their autonomy. Floridi argues that if such a system were to gain sufficient credibility, we would consider it as defining standards.

What is Success?

This search revealed research groups that successfully created and routinely used intelligent systems. This routine is not the only indicator of success. An entire area of published research, Information System (IS) Success, is devoted to establishing criteria by which an IS can be judged to be a success or failure. However, selecting which of these criteria is essential is not straightforward. DeLone and McLean reviewed 180 studies showing successful weather systems and found that those 180 studies used over 100 different variables (Wu & Wang, 2006). The authors can measure success. These authors included six factors in the model: a system's quality, information quality, usage of information technology (IT), user satisfaction, individual impact, and organizational effect (Gorla et al., 2010). DeLone and McLean's original model was too complex and complicated for Shim & Jo (2020), so the authors came up with a model of success based on just five variables: system quality, perceived utility, user happiness, and the usage of information systems.

Research in Information Systems Success is well-versed in DeLone, McLean, and Seddon's models. Jeyaraj (2020) evaluated how well these models measured IS success and concluded they were both helpful. Their finding shows that either model can be applied to rich data to establish whether an IS can be considered a success or not. However, at the stage in this research when it was necessary to select two case studies for in-depth examination, information about factors such as system quality and perceived usefulness was not reliably available for all systems created by the 58 research groups that had responded to the questions about their designs. It was impossible to establish rigorously from the data available which research groups had thus created the most "successful" systems using models such as the above. For this reason, more specific criteria for case study selection were required and used.

From the data available, it was possible to establish how many systems each research group had created and brought into routine use. IS uses the variable selected for both Information systems. Regardless of origin, they must be integrated into the business process and assimilated into the social fabric of their host company. This integration showed that these authors, too, thought it was a key factor. Besides the direct relationship between a system being in routine use and the 'IS use' variable of Seddon's model, there is an implied relationship between three others of Seddon's variables. If a system is in routine use, it would have to show reasonably high levels of user happiness. System and data quality are intertwined here. Personal and organizational ramifications would increase if IS were implemented. DeLone and McLean's other two variables. As a side effect of this research, these expectations were confirmed: the systems that were looked at in-depth met these criteria.

Further, if a group had many systems in routine use, there would be a concentration of data in one location. More people would be involved in the systems' development, testing, or use. For a group to create many designs, they probably must have encountered many problems in the past that they must have successfully overcome to reach their current status. Both from the ISI Citation database, http://wok.mimas.ac.uk variables, would increase the richness of the data that could be gathered to solve the research questions. Because of this, we chose groups that used as many systems as possible every day so that as much data as possible could be collected in the time we had.

SOLUTIONS AND RECOMMENDATIONS

It is possible that any researcher who has successfully created and routinely used intelligent systems would have had to overcome some, or perhaps these, difficulties.

We decided that the next part of the research should examine the challenges the two most successful groups had faced and how they overcame them. "Most successful" was considered to be those groups who had created the most systems used routinely in medicine. We felt that a group with more designs in routine use would likely have more success stories to relate to and a better understanding of how they had achieved their success. We thought that publishing this information could help less-successful groups meet these challenges. This publication may encourage the successful production and use of other intelligent systems in medicine.

FUTURE RESEARCH DIRECTIONS

The empirical work must be conducted correctly, and sufficient analysis of the data collected in the practical work should be carried out in a possible course of study for the future.

CONCLUSION

Intelligent systems, an artificially active healthcare approach, are expected to grow in the future. Ethical considerations of intelligent systems in medicine are addressed in this chapter. This section summarises four ethical considerations. Accuracy, accessibility, privacy, and ownership are the core values, and any use of intelligent systems in medicine must comply with these ethical considerations.

REFERENCES

Alazzam, M. B., Tayyib, N., Alshawwa, S. Z., & Ahmed, M. (2022). Nursing care systematization with case-based reasoning and artificial intelligence. *Journal of Healthcare Engineering*, *2022*, 2022. doi:10.1155/2022/1959371 PMID:35310193

Askitopoulou, H., & Vgontzas, A. N. (2018). The relevance of the Hippocratic Oath to the ethical and moral values of contemporary medicine. Part I: The Hippocratic Oath from antiquity to modern times. *European Spine Journal*, *27*(7), 1481–1490. doi:10.100700586-017-5348-4 PMID:29080001

Brewer, E., Demmer, M., Du, B., Ho, M., Kam, M., Nedevschi, S., Pal, J., Patra, R., Surana, S., & Fall, K. (2005). The case for technology in developing regions. *Computer*, *38*(6), 25–38. doi:10.1109/MC.2005.204

Brown, K. C. (2018). *A Consideration of Mason's Ethical Framework: The Importance of PAPA Factors in the 21st Century: A Seven-Year Study*. University of Florida.

Cai, C. J., Winter, S., Steiner, D., Wilcox, L., & Terry, M. (2019). "Hello AI": uncovering the onboarding needs of medical practitioners for human-AI collaborative decision-making. *Proceedings of the ACM on Human-computer Interaction*, *3*(CSCW), 1-24. 10.1145/3359206

Cockburn, I. M., Henderson, R., & Stern, S. (2018). The impact of artificial intelligence on innovation: An exploratory analysis. In *The economics of artificial intelligence: An agenda* (pp. 115–146). University of Chicago Press.

Coventry, L., & Branley, D. (2018). Cybersecurity in healthcare: A narrative review of trends, threats, and ways forward. *Maturitas*, *113*, 48–52. doi:10.1016/j.maturitas.2018.04.008 PMID:29903648

Dilsizian, S. E., & Siegel, E. L. (2014). Artificial intelligence in medicine and cardiac imaging: Harnessing big data and advanced computing to provide personalized medical diagnosis and treatment. *Current Cardiology Reports*, *16*(1), 1–8. doi:10.100711886-013-0441-8 PMID:24338557

Ekblaw, A., Azaria, A., Halamka, J. D., & Lippman, A. (2016, August). A Case Study for Blockchain in Healthcare: "MedRec" prototype for electronic health records and medical research data. In *Proceedings of IEEE open & big data conference* (Vol. 13, p. 13).

Floridi, L. (Ed.). (2010). *The Cambridge handbook of information and computer ethics*. Cambridge University Press. doi:10.1017/CBO9780511845239

Glasgow, R. E., Vinson, C., Chambers, D., Khoury, M. J., Kaplan, R. M., & Hunter, C. (2012). National Institutes of Health approaches to dissemination and implementation science: Current and future directions. *American Journal of Public Health*, *102*(7), 1274–1281. doi:10.2105/AJPH.2012.300755 PMID:22594758

Gorla, N., Somers, T. M., & Wong, B. (2010). Organizational impact of system quality, information quality, and service quality. *The Journal of Strategic Information Systems*, *19*(3), 207–228. doi:10.1016/j.jsis.2010.05.001

Jeyaraj, A. (2020). DeLone & McLean models of information system success: Critical meta-review and research directions. *International Journal of Information Management*, *54*, 102139. doi:10.1016/j.ijinfomgt.2020.102139

Kaelber, D. C., Jha, A. K., Johnston, D., Middleton, B., & Bates, D. W. (2008). A research agenda for personal health records (PHRs). *Journal of the American Medical Informatics Association*, *15*(6), 729–736. doi:10.1197/jamia.M2547 PMID:18756002

Kaur, P. D., & Chana, I. (2014). Cloud-based intelligent system for delivering health care as a service. *Computer Methods and Programs in Biomedicine*, *113*(1), 346–359. doi:10.1016/j.cmpb.2013.09.013 PMID:24139021

Keshta, I., & Odeh, A. (2021). Security and privacy of electronic health records: Concerns and challenges. *Egyptian Informatics Journal*, *22*(2), 177–183. doi:10.1016/j.eij.2020.07.003

Khan, I., Dewangan, B., Meena, A., & Birthare, M. (2020, March). Study of Various Cloud Service Providers: A Comparative Analysis. At *the 5th International Conference on Next Generation Computing Technologies*, 1-5. 10.2139srn.3672950

Krittanawong, C., Johnson, K. W., Choi, E., Kaplin, S., Venner, E., Murugan, M., & Tang, W. W. (2022). Artificial Intelligence and Cardiovascular Genetics. *Life (Chicago, Ill.)*, *12*(2), 279. PMID:35207566

Li, M., & Chapman, G. B. (2020). Medical decision-making. The Wiley Encyclopedia of Health Psychology, 347-353.

Longoni, C., Bonezzi, A., & Morewedge, C. K. (2019). Resistance to artificial medical intelligence. *The Journal of Consumer Research*, *46*(4), 629–650. doi:10.1093/jcr/ucz013

Luengo-Oroz, M., Hoffmann Pham, K., Bullock, J., Kirkpatrick, R., Luccioni, A., Rubel, S., Wachholz, C., Chakchouk, M., Biggs, P., Nguyen, T., Purnat, T., & Mariano, B. (2020). Artificial intelligence cooperation to support the global response to COVID-19. *Nature Machine Intelligence*, *2*(6), 295–297. doi:10.103842256-020-0184-3

Maedche, A., Legner, C., Benlian, A., Berger, B., Gimpel, H., Hess, T., Hinz, O., Morana, S., & Söllner, M. (2019). AI-based digital assistants. *Business & Information Systems Engineering*, *61*(4), 535–544. doi:10.100712599-019-00600-8

Mason, R. O. (2017). Four ethical issues of the information age. In *Computer ethics* (pp. 41–48). Routledge. doi:10.4324/9781315259697-8

Morley, J., Floridi, L., Kinsey, L., & Elhalal, A. (2020). From what to how: An initial review of publicly available AI ethics tools, methods, and research to translate principles into practices. *Science and Engineering Ethics*, *26*(4), 2141–2168. doi:10.100711948-019-00165-5 PMID:31828533

Musen, M. A., Middleton, B., & Greenes, R. A. (2021). Clinical decision-support systems. In *Biomedical informatics* (pp. 795–840). Springer. doi:10.1007/978-3-030-58721-5_24

Nissenbaum, H. (2020). Protecting privacy in an information age: The problem of privacy in public. In *The Ethics of Information Technologies* (pp. 141–178). Routledge. doi:10.4324/9781003075011-12

Osoba, O. A., & Welser, I. V. W. (2017). *An intelligence in our image: The risks of bias and errors in artificial intelligence.* Rand Corporation.

Shim, M., & Jo, H. S. (2020). What quality factors affect the perceived benefits of online health information sites? Application of the updated DeLone and McLean Information Systems Success Model. *International Journal of Medical Informatics, 137*, 104093. doi:10.1016/j.ijmedinf.2020.104093 PMID:32078918

Sparrow, R. (2007). Killer robots. *Journal of Applied Philosophy*, *24*(1), 62–77. doi:10.1111/j.1468-5930.2007.00346.x

Toreini, E., Aitken, M., Coopamootoo, K., Elliott, K., Zelaya, C. G., & Van Moorsel, A. (2020, January). The relationship between trust in AI and trustworthy machine learning technologies. In *Proceedings of the 2020 conference on fairness, accountability, and transparency* (pp. 272-283). 10.1145/3351095.3372834

Veatch, R. M. (2012). *Hippocratic, religious, and secular medical ethics: the points of conflict.* Georgetown University Press.

Ventola, C. L. (2014). Social media and health care professionals: Benefits, risks, and best practices. *P&T*, *39*(7), 491. PMID:25083128

Wu, J. H., & Wang, Y. M. (2006). Measuring KMS success: A respecification of the DeLone and McLean's model. *Information & Management*, *43*(6), 728–739. doi:10.1016/j.im.2006.05.002

Yaacoub, J. P. A., Noura, M., Noura, H. N., Salman, O., Yaacoub, E., Couturier, R., & Chehab, A. (2020). Securing the Internet of medical systems: Limitations, issues, and recommendations. *Future Generation Computer Systems*, *105*, 581–606. doi:10.1016/j.future.2019.12.028

Young, J. A., Smith, T. J., & Zheng, S. H. (2020). Call me BIG PAPA: An extension of Mason's information ethics framework to big data. *Journal of the Midwest Association for Information Systems*, *2*, 17–41.

ADDITIONAL READING

Komalasari, R. (2022). A Social Ecological Model (SEM) to Manage Methadone Programmes in Prisons. In Handbook of Research on Mathematical Modeling for Smart Healthcare Systems (pp. 374-382). IGI Global.

Komalasari, R. (2022). Pemanfaatan Kecerdasan Buatan (Ai) Dalam Telemedicine: Dari Perspektif Profesional Kesehatan. *Jurnal Kedokteran Mulawarman*, *9*(2), 72–81.

Komalasari, R. (2023). Telemedicine in Pandemic Times in Indonesia: Healthcare Professional's Perspective. In *Health Informatics and Patient Safety in Times of Crisis*. IGI Global.

Komalasari, R. (in press). Digital twin elderly healthcare services. In *Digital Twins and Healthcare: Trends, Techniques, and Challenges*. IGI Global.

Komalasari, R. (in press). Spirituality and health benefits. In *Supporting Children and Youth Through Spiritual Education*. IGI Global.

Komalasari, R. (in press). History and Legislative Changes Governing Medical Cannabis in Indonesia. In *Medical Cannabis and the Effects of Cannabinoids on Fighting Cancer, Multiple Sclerosis, Epilepsy, Parkinsons and Other Neurodegenerative Diseases*. IGI Global.

Komalasari, R. (in press). Designing Health Systems for Better, Faster, and Less Expensive Treatment. In *Exploring the Convergence of Computer and Medical Science Through Cloud Healthcare*. IGI Global. doi:10.4018/978-1-6684-5260-8.ch001

Komalasari, R. (in press). Designing Health Systems for Better, Faster, and Less Expensive Treatment. In *Exploring the Convergence of Computer and Medical Science Through Cloud Healthcare*. IGI Global. doi:10.4018/978-1-6684-5260-8.ch001

Komalasari, R. (in press). Treatment of menstrual discomfort in young women and a cognitive behavior therapy (CBT) program. In *Perspectives on Coping Strategies for Menstrual and Premenstrual Distress, has just been completed*. IGI Global.

Komalasari, R. (2022). Literature Review: Health Aspects and Legal Protection for Children in Scotland and Indonesia. *Buana Gender: Jurnal Studi Gender dan Anak*.

Komalasari, R. (2022). Manfaat Positif Allium Sativum L. (Bawang Putih) Dalam Kaitannya Dengan Berbagai Penyakit. [JFM]. *Jurnal Farmasimed*.

Komalasari, R. (2022). Pelatihan Kesehatan dan Pengembangan Profesional Berkelanjutan untuk Pendidik Olah Raga. *Jurnal Ilmiah STOK Bina Guna Medan*.

Komalasari, R. (2022). Persepsi Hakim tentang Rehabilitasi Pengguna Narkoba:Tantangan dan Peluang. *Arena Hukum*.

Komalasari, R., & Mustafa, C. (2021). Meningkatkan Pelayanan Administrasi Publik di Indonesia. *PaKMas: Jurnal Pengabdian Kepada Masyarakat*, *1*(1), 20–27. doi:10.54259/pakmas.v1i1.29

Komalasari, R., & Mustafa, C. (2021). Pendidikan Profesi dan Pengabdian Masyarakat di Indonesia. *PaKMas: Jurnal Pengabdian Kepada Masyarakat*, *1*(1), 28–36. doi:10.54259/pakmas.v1i1.30

Komalasari, R., & Mustafa, C. (2022). Empowerment of Women with Narcotic Cases. *Buana Gender: Jurnal Studi Gender dan Anak*, 7(1).

Komalasari, R., Nurhayati, N., & Mustafa, C. (2022). Insider/Outsider Issues: Reflections on Qualitative Research. *Qualitative Report*, *27*(3), 744–751. doi:10.46743/2160-3715/2022.5259

Komalasari, R., Nurhayati, N., & Mustafa, C. (2022). Enhancing the Online Learning Environment for Medical Education: Lessons From COVID-19. In Policies and procedures for the implementation of safe and healthy educational environments: Post-COVID-19 perspectives (pp. 138-154). IGI Global.

Komalasari, R., Nurhayati, N., & Mustafa, C. (2022). Kebijakan Penanganan Penyintas HIV/AIDS Di Lembaga Pemasyarakatan. *Jurnal Kesehatan Kartika*, *17*(1), 19–27.

Komalasari, R., Nurhayati, N., & Mustafa, C. (2022). Professional Education and Training in Indonesia. In *Public Affairs Education and Training in the 21st Century* (pp. 125–138). IGI Global. doi:10.4018/978-1-7998-8243-5.ch008

Komalasari, R., Wilson, S., & Haw, S. (2021). A systematic review of qualitative evidence on barriers to and facilitators of the implementation of opioid agonist treatment (OAT) programmes in prisons. *The International Journal on Drug Policy*, *87*, 102978. doi:10.1016/j.drugpo.2020.102978 PMID:33129135

Komalasari, R., Wilson, S., & Haw, S. (2021). A social ecological model (SEM) to exploring barriers of and facilitators to the implementation of opioid agonist treatment (OAT) programmes in prisons. *International Journal of Prisoner Health*, *17*(4), 477–496. doi:10.1108/IJPH-04-2020-0020

Komalasari, R., Wilson, S., Nasir, S., & Haw, S. (2020). Multiple burdens of stigma for prisoners participating in Opioid Antagonist Treatment (OAT) programmes in Indonesian prisons: A qualitative study. *International Journal of Prisoner Health*.

Mustafa, C. (2021). The Challenges to Improving Public Services and Judicial Operations: A unique balance between pursuing justice and public service in Indonesia. In Handbook of research on global challenges for improving public services and government operations (pp. 117-132). IGI Global. doi:10.4018/978-1-7998-4978-0.ch007

Mustafa, C. (2021). Key finding: result of a qualitative study of judicial perspectives on the sentencing of minor drug offenders in Indonesia: Structural inequality. *Qualitative Report*, *26*(5), 1678–1692. doi:10.46743/2160-3715/2021.4436

Mustafa, C. (2021). The view of judicial activism and public legitimacy. *Crime, Law, and Social Change*, *76*(1), 23–34. doi:10.100710611-021-09955-0

Mustafa, C. (2021). Qualitative method used in researching the judiciary: Quality assurance steps to enhance the validity and reliability of the findings. *Qualitative Report*, *26*(1), 176–186. doi:10.46743/2160-3715/2021.4319

Mustafa, C. (2021). The News Media Representation of Acts of Mass Violence in Indonesia. In Mitigating Mass Violence and Managing Threats in Contemporary Society (pp. 127-140). IGI Global. doi:10.4018/978-1-7998-4957-5.ch008

Mustafa, C., Malloch, M., & Hamilton Smith, N. (2020). Judicial perspectives on the sentencing of minor drug offenders in Indonesia: Discretionary practice and compassionate approaches. *Crime, Law, and Social Change*, *74*(3), 297–313. doi:10.100710611-020-09896-0

Suhariyanto, B., Mustafa, C., & Santoso, T. (2021). Liability incorporate between transnational corruption cases Indonesia and the United States of America. *Journal of Legal, Ethical and Regulatory Issues*, *24*(3).

Suhariyanto, B., & Mustafa, C. (2022). Analysis And Evaluation Of Legal Aid In The Indonesian Court. *Jurnal Hukum dan Peradilan*, 11(2), 176-194.

KEY TERMS AND DEFINITIONS

Accessibility: As computers increasingly take over business record-keeping tasks, the accuracy or correctness of information gains actual significance for society.

Accuracy: It has been described in terms of people's capacity to actively manage personal data. But making people responsible for their own information would dramatically restrict what can be considered private.

Ownership: These rights raise significant economic and ethical questions, questions centered on the unique characteristics of information itself and the channels through which it is disseminated.

Privacy: Fundamentally, it is about safeguards against unauthorized access and information collecting. It has typically been characterized in terms of people's capacity to exercise personal control over information about themselves.

Chapter 2
Advanced Bioinspiration Methods for Optimisation Problems:
Survey and Applications

Hadj Ahmed Bouarara
https://orcid.org/0000-0002-4973-4385
Dr. Moulay Tahar University, Algeria

ABSTRACT

Over the past decade, solving complex optimization problems with metaheuristics has received considerable attention among practitioners and researchers. As a result, many metaheuristic algorithms have been developed in recent years, and a large majority of these algorithms are often inspired by nature. Today, bio-inspired methods are becoming more and more popular. This popularity and success stems mainly from the fact that these algorithms were developed by mimicking nature's most efficient processes, including biological systems and physical and chemical processes. In this chapter, we will talk about metaheuristics and bio-inspired algorithms in general. From their basic concepts to their applications.

1. INTRODUCTION AND BACKGROUND

The algorithms play an important role in IT and for several real applications. It delivers an objective is to find answers, which are probably best likened to the runtime environment. In late years, engineers and decision makers are confronted daily with complex problems (NP-hard) that generally involve all sectors. Historically,

DOI: 10.4018/978-1-6684-5656-9.ch002

researchers have attempted to resolve these problems as efficiently as possible. For many years ago, research has been conducted towards the proposition of exact algorithms for polynomial special problems. Afterwards, the appearance of heuristic algorithms allowed generally finding solutions with good quality but often solutions for small instances, so why the need to find new types of algorithms that can lead to a major breakthrough for the practical resolution these problems became paramount.

Today, a huge success was achieved through modelling of organic and natural intelligence resulting in what is called "computational intelligence algorithms". This class of algorithms (include artificial neural networks, evolutionary computations, collective intelligence, artificial immune systems, human organ systems and fuzzy systems) constitutes a part of meta-heuristic and bio-mimicry areas. They have demonstrated their strength face to different complex issues where they are even attempting to determine the optimal solution from a finite number of existing solutions and offer a high performance results in experimental studies. It is frequently hard to understand why they perform well in a particular context. Another significant advantage is that these algorithms can often be applied without much knowledge about the problem, which makes them very suitable for various applications.

The conception of such algorithm requires the presence of the three characteristics to facilitate the implementation of these algorithms on a new problem:

- Choose a representation of possible solutions.
- Determine a function to measure the quality of a solution.
- Define the operators producing from a current set of solutions a new set of solutions.

It is no overstatement to state that this type of algorithm is everywhere, from design engineering to business planning and from routing of network to travel planning. In all these actions, we strain to reach some goals or optimize something like the quality of performance and the execution time. The delegation of this paper is a very important way to consolidate a number of new algorithms inspired by nature and have been offered in the literature. It is composed from more than 20 algorithms that were unionised in 2 parts classified by the biological source of inspiration of each one of them as illustrated in the next figure 1.

1.2 The Organization of Our Work

We describe the algorithm following the next structure:

- **The Origin of the Algorithm**

Figure 1. Taxonomy of different bio-inspired algorithms existed in literature

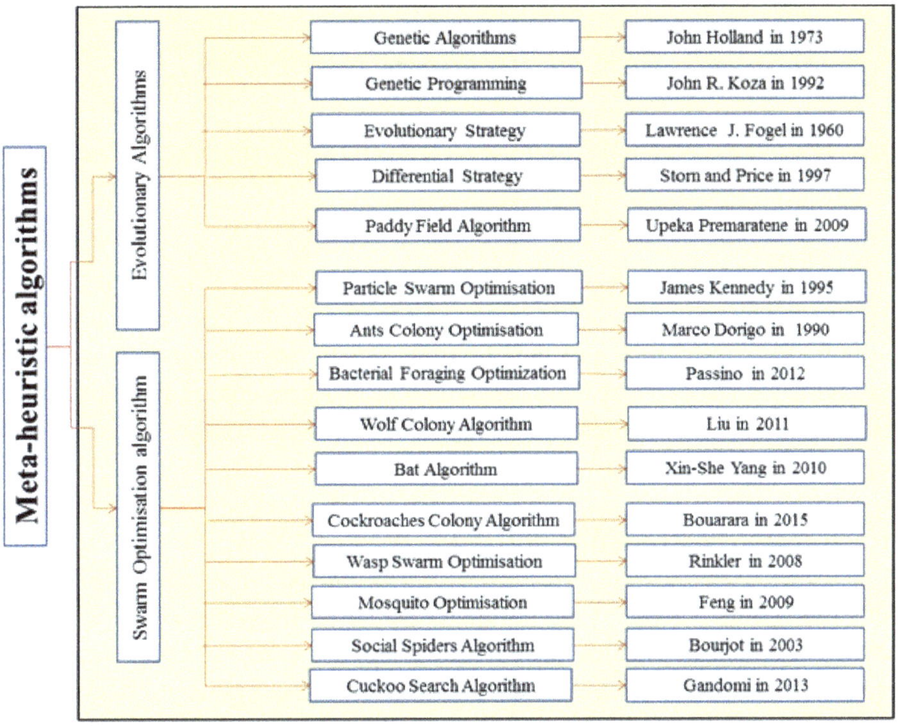

It describes the start and the first appearance of this algorithm with references regarding the creator or developer of the algorithm.

- **The Inspiration Source:**

It traces the inspiration phenomenon of the algorithm that can be natural, biological, physical or social.

- **The Summary of the Algorithm:**

It summarizes the general cognitive process and the primary goal of the algorithm.

- **The General Process of the Algorithm:**

It describes in more detail the various steps of the algorithm and the architecture, in the form of the figure.

Figure 2. The process (problem / solution) for an evolutionary algorithm

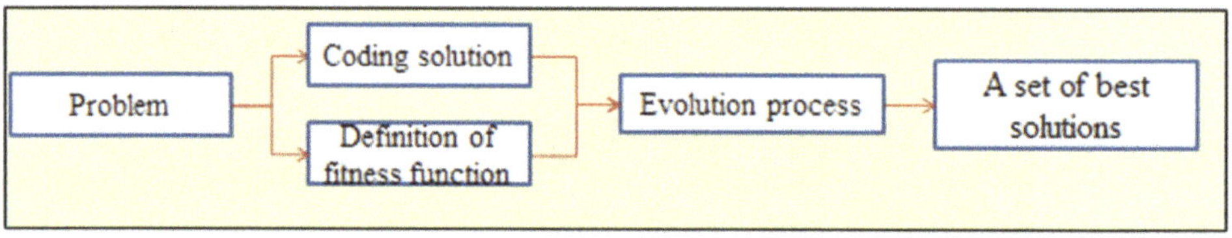

- **The Procedure:**

It is the structured image of the algorithm which describes its arrangement in terms of the necessary parameters, the input/output, the stopping criterion, and the intercourse between the different calculations performed by the constituents of the algorithm.

2. Evolutionary Algorithms (EA)

In the 1950s until 1960, several computer scientists have studied independently scalable systems with the idea that evolution could be used as an optimization tool for engineering problems. Evolutionary algorithm (EA) is a family of stochastic and probabilistic algorithms that mimic the metaphor of biological evolution to solve several known problems in the computer area. These algorithms are established on the mechanisms of the theory of evolution proposed by Darwin in 1860 and published in his book "The Origin of Species by means of Natural Selection, or the struggle for existence in nature" (Darwin, 1859). They accept as input a population of individuals (set of potential answers) that will evolve from one generation to another following a set of reproduction rules (selection, mutation, crossing, and replacement) until the optimal solution satisfies the targets set in advance (stopping criterion). This process leads to the evolution of a population of individuals that are better adapted to their environment; as, the natural life, iteratively a new population of individuals will be created (Pohlheim, 2005).

2.1.1. The General Process of Evolution

All algorithms belonging to the EA family follow the same principle (problem / solution) present in Figure 2 where the only difference between them lies in the type of coding solutions used and the techniques employed in breeding rules.

The functioning of the evolutionary process for EA consists of 5 following steps:

- **Initialisation:** We start with an initial population of potential solutions (chromosomes) arbitrarily chosen.
- **The coding of individuals.**
- **Evaluation:** We evaluate the performance (fitness value) of each individual.
- **Reproduction operators:** Create a new population of potential solutions using simple evolutionary operators of reproduction such as selection, crossover and mutation.
- **Replacement:** We restart this cycle until we discover a satisfactory solution (stopping criterion).

2.1.2. The General Structure of Evolutionary Algorithms:

The general structure of an evolutionary algorithm is grouped in the following pseudo code:

```
The General procedure of evolutionary algorithm
Input: PC: crossover probability; PM: mutation probability; MI:
best individual; cd; stopping rule
Encoding (P (i))
Begin
     I          0
    Randomly-generated (P (i))
    Evaluate (P (i))
    While (no stopping criterion) do
       Begin
            i           i + 1
           Select P (i) from P (i-1)
           Cross (P (i))
           Mutate (P (i))
           Mi              evaluate (P (i))
           P (i)                 Replace (P (i))
       End
   End
Output: Mi
```

The EA are computationally expensive and do not guarantee the optimality of the results where it is difficult to implement them. Among all the EAs existed in literature, we have found:

2.2. Genetic Algorithms (GAs)

2.2.1. The Origin of the Algorithm

The commencement of genetic algorithms can be delineated to the early 1950s, when several biologists use computers to simulate biological systems (Goldberg, 1989). Nevertheless, in the 1960s, John Holland studied the evolutionary systems during his doctorate at the University of Michigan, USA and in 1975 he introduced the first formal model of GAs called the canonical genetic algorithms (CGA) in his book "adaptation in natural and artificial systems" (Holland, 1962; Holland, 1973).

2.2.2. The Inspiration Source

The GAs is a meta-heuristic algorithm optimization based on the principle of Neo-Darwinism derived from the Mendelian genetics (including heredity, the frequency of genes, chromosomes and alleles) and the natural evolution of Darwin mechanisms such as crossing, mutation, and natural selection (Holland, 1975).

2.2.3. The Summary of the Algorithm

The GAs is the best known evolutionary algorithms that provide solutions to problems that do not have calculable solutions in reasonable time, analytically or algorithmically. They accept as input a population of N solutions represented by individuals (named the parent population) that will evolve iteratively using genetic operators based on a fitness function to get the best solutions. The foundation of a new generation is achieved by creating new individual and destroying other (natural selection mechanism), allowing the renewal of the population and for each generation only a subset of individuals is able to reproduce and transmit the characters to the future generations. The original goal of Holland was not to design algorithms to solve specific problems, but rather to formally study the phenomenon of adaptation as it occurs in nature and to develop means by which natural coping mechanisms could be imported into computer systems (Reeves, 2003).

2.2.4. The General Process

The universal architecture of genetic algorithms is shown in Figure 2 (Section 2.1.1) and follows the same mathematical process as the process demonstrated in the old part. Each step in the functioning of GAs is detailed in the next:

2.2.4.1. Initialization

The foremost stride in the implementation of GAs is to create an initial population of individuals that should contain chromosomes distributed in all search space. The easiest way is to randomly generate N individuals. This step can condition the speed of the algorithm and the quality of the outcome.

2.2.4.2. Coding

The encoding is a delicate and vital step in designing a genetic algorithm because it permits the representation of individual in the contour of a chromosome for the carrying out of the reproduction operators. There are three types of coding techniques: binary, real and grey coding (Mitchell, 1998).

2.2.4.3. The Evaluation Function

The evaluation function (fitness) allows to measure the performance of each individual and to evaluate whether an individual is better suited than another to its environment. This feature, directly related to the problems is often simple to define it when there are few parameters. Unluckily, when there are lots of objectives, it becomes more difficult to fix it. In this case, the function is a weighted sum of several functions (multi-objective function).

2.2.4.4. Reproduction Operators

These operators play a key role in the success of AGs that can assure the passage from one generation to another in order to improve the population:

- **The Selection Process:**

Natural selection proposed by Darwin (1859) and Wallace (1870) in the nineteenth century, is based on the principle that individuals tend to make more children and cannot be reproduced all (Wallace, 1870). This operator may be the most important because it allows to individuals in a population to survive, reproduce or die (identify the individuals to be crossed). In general, the probability of survival of an individual will be directly connected to its relative effectiveness within the population and the best individuals can be selected several times for the next generation, while the less able will have less chance of being.

- **Crossover Operator:**

The crossover operator allows the institution of new individuals through exchanging of information between chromosomes (individuals). A cross is envisaged with two parents selected and generates one or two children who inherit some characteristic of their parents. To do this task we need a crossover probability fixed in advance MP after selecting two individuals we generate a random number $A \in [0, 1]$.If $A \leq PC$ then a crossing will be done along these two parents individuals. The intersection zone is generally chosen randomly. We can distinguish several techniques to realise the crossing.

- The cross in a point:

Both parents will be separated into two sections with a single break point. The cutoff position is randomly chosen and will be the same for both parents to transmit an information exchange.

- The cross in two points:

Both parents will be divided into 3 parts with two break points. The cutoff areas will be randomly selected and will be the same for both parents to perform the interchange of data between them.

- Uniform crosses:

It is based on the same principle of previous cross types, but the points of the cross can be multiple and randomly chosen.

- **Mutation Operator:**

It is an operator that provides exploration of the search space and ensures the diversity of the population with the general principle is to change a gene in a chromosome to another. For instance, in the binary encoding, we change a bit with its opposite. This is applied with a mutation probability (MP) fixed in advance which has usually a very low value between 0.001 and 0.01. We must randomly choose a variable A and if this variable is less than MP then the mutation will realized.

- **Replacement:**

This operator is to reintroduce children obtained by the successive application of the production operators (selection, crossover and mutation) on the population of the iteration me into the population of their parents (the population of iteration $I + 1$).

2.2.4.5. The Procedure

The general structure of genetic algorithms is grouped in the following pseudo code:

```
Genetic Algorithms (GAs)
CP: crossover probability; MP: mutation probability, CA:
stopping criterion; Ni: the size of the population P; A: real
variable; Mi: best individual (solution).
Input: PC, PM, CA, Ni, A
Begin
    Initialization and encoding of the initial population P:
   Evaluate (P)
    While (No CA) do
      P '= Selection of Parents in P
      If rand (A) <PC then P '=Apply Crossing operator (P ')
      If rand (A) <PM then P '= Apply Mutation operator (P ')
      P                              Replace the elders of P by
their Descendants P '.
     Mi              Evaluate (P)
   End while
End
Output: Mi
```

2.3. Genetic Programming (GP)

2.3.1. The Origin of the Algorithm

The GP was proposed by John Koza in (Koza, 1992), as a particular class of evolutionary algorithms used for the development of computer programs. Koza has developed the GP by combining between the study released by Cramer in 1985 on the wooded-health structure for the representation of individual programs as well as the crossing under the parent trees (Cramer, 1985) and the study by Hicklin in 1986 on selecting and combining under trees (Hicklin, 1986).

2.3.2. The Inspiration

The Genetic Programming has the same inspiration source as genetic algorithms (selection mechanism and natural evolution) but it concentrates a lot along the evolution of genotypes. It has been handled as a subset of genetic algorithms, but the conflict is that the genotype is actually presented as a tree rather than a string or vector.

2.3.3. Summary of the Algorithm

Genetic programming (GP) is the last algorithm from the AE family that have emerged. It can automatically solve problems without the specification of the form or construction of the solution in advance by the user. At the most abstract level, GP is an independent systematic approach in the sphere because it provides information processing systems that can resolve problems automatically from a high-level statement of what must be done. Since its origin, the GP has attracted the interest of the myriad of people round the globe.

The objective of the GP is to use induction to design a computer program with the principle of evolving a set of computer programs from one generation to another with a tree structure for coding programs (individuals) to improve the fit between the candidate population programs and an objective function. The GP is a random procedure that does not guarantee the optimality of the results (Poole, 2008).

2.3.4. The General Process

The basic steps of GP are indicated in the next Figure 3, initially it discovers each program of the initial population by running and comparing their behaviour with another ideal. We might be interested, for example, how a program includes a series of time or controls an industrial operation. This comparison is carried out using a fitness function (evaluation) set in advance according to the desired objective. The programs with the best fitness function (parent programs) will be selected to breed and produce new programs (children's program) for the next generation using the primary genetic operations (selection, crossover and mutation) (Koza, 1994).

2.3.4.1. Initial Programs Population

Like other evolutionary algorithms, in GP the programs of the first population must be generated randomly. There's a number of different approaches to generate this population as the global method and ramped method (KOZA 1992).

2.3.4.2. Coding Programs

In GP, each individual is a computer program represented as a nested list, which can be directly mapped into a syntax tree constituting from:

- Variables and constants that represent the leaves of the tree or terminals.

Figure 3. The general functioning of genetic programming algorithm

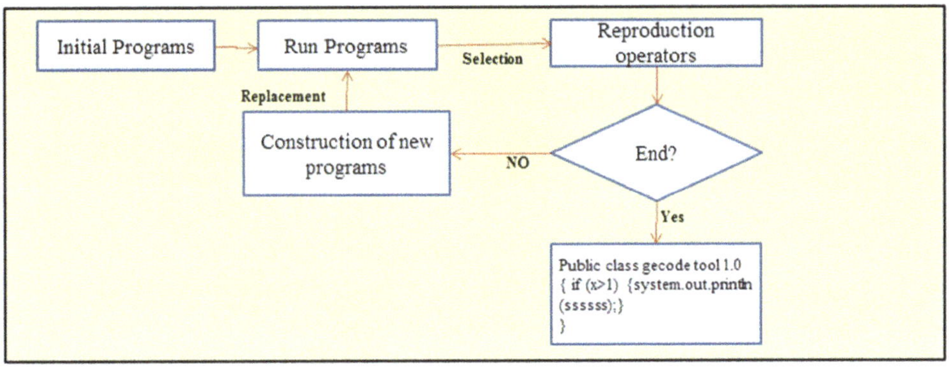

- Arithmetic operations are called internal nodes function.
- The functions and terminals from the primitive set of a PG system.

2.3.4.3. The Selection Operator

Genetic operators GP are applied to programs that are selected according to their fitness. In other words the best programs are liable to have more children programs. The most used selection method by GP, is the tournament technique (for more details see Section 2.2.4).

2.3.4.4. Reproduction Operators

PG differs significantly from other evolutionary algorithms in the implementation of crossover and mutation operators.

- **The crossover Operator:**

The creation of new children programs by combining two random parts from two sub trees of selected parent programs. For this purpose a method called the cross in a sub-tree structure is used.

- **The Mutation Operator:**

Creating a new child program, by modifying randomly a selected part of a sub father tree. The replacement step in GP will remain the same as the replacement method of GAs (more detail in section 2.2.4).

2.3.4.5. The Procedure

The general structure of the GP is grouped in the next pseudo code:

```
Genetic programming procedure
AM: Random Variable crossover [0-1] MP: best program; MP:
mutation probability; CP: cross probability; AM: random
mutation variable [0-1]; AM: Random Variable crossover [0-1],
Pi{}: size list of Npg programs initially generated from the
set of primitives available, NG{}: set of programs (parent and
son) that will be built after each generation, Npg: the size
of the list Pi{} and NG{}, PGi: program number I of population
programs, Sp{}: the list of selected programs, SPGi: program
number I from the selected programs.
Input: CP, MP, Pi
For each PGi from Pi do
        Execute and evaluate the program i.
end
 while (no sttoping cretirien) do
                                *selection operator*
                 SP                               tournament-
selection (PGi) (for more detail see the section 2.2.4)
                 *The cross operator*
        For each (PGi and PGj) selected randomly from Sp do
             Generat-randomly (AC)
               If AC<PC then NG{}
interbreed (SPGi, SPGj)  (i must be different to j)
               else NG{}                        do-not-
cross(SPGi, SPGj)
        end
                                   *mutation operator*
        For each PGi from NG do
            Generate-randomly(AM)
            if AM<PM then NG{}
mutation(SPGi)
            else do-not-mutation(SPGi)
        end
      for each PGi from NG do evaluate (PGi)
           Mp                                   best of (PGi)
```

```
end
output: M_p
```

2.4. Evolutionary Strategy (ES)

2.4.1. The Origin of the Algorithm (ES)

Evolutionary Strategy (SE) is chronologically the first of all the evolutionary algorithms. It dates back to the 1960s and is the most commonly applied as a black box in optimization problems in continuous search spaces, discrete and also in the case of the multi-objective constraints. The algorithm of the SE was born in Germany in the 60s and the first work was presented in (Fogel, 1962) to solve numerical optimization problems. It was later developed by Ingo Rechenberg, Hans-Paul Schwefel and colleagues in 1964 to resolve an aerodynamic design problem (Rechenberg, 1989).

2.4.2. The Inspiration

Evolutionary strategy is an optimization technique based on ideas of adaptation and evolution. It is animated by the theory of adaptation and evolution by natural selection and the macro level in the process of species evolution (phenotype, hereditary variation) unlike the genetic algorithm that trait the micro level at the genomic (genomic, chromosomes, genes, alleles).

2.4.3. The Summary of the Algorithm

The ES algorithm called evolutionary programming is particularly a suitable paradigm for nonlinear optimization spots. Like other evolutionary algorithms, genetic operators are applied in a loop. An iteration of the loop is called a generation. The sequence of generations is continued until a stopping criterion is reached. A very important feature of SE is the use of self-adaptive mechanisms to control the application of a mutation. These mechanisms are designed to optimize the research progress, by changing not only the solutions to the problem to study, but also some settings to mutate these solutions.

2.4.4. The General Process:

The algorithm of evolution strategies is a specific instance of evolutionary algorithms and its operation is characterized by the following four properties:

- The selection of individuals for recombination is impartial, that is based solely on fitness rankings.
- The selection is a deterministic process.
- The mutation operator is parameterized and therefore it can change its properties during optimization.
- Individuals consist of decision parameters and Strategy Settings.

The operation of this algorithm is based on coding individuals, usually by a vector where each individual component of the vector is a real number, a deterministic selection, a cross (discreet for the parameters of the strategy and means for decision parameters) and mutation for the strategy and for the decision. Some forms of selection and sampling in the SE are as follows:

- SE - (1 + 1): This is a simple selection mechanism in which works by creating a vector of real value of the object variable from its parents by applying a mutation with a standard deviation identical to each variable object. Then, the individual results are evaluated and compared to their parents and if it has a greater fitness relative to those parents then it becomes a parent for the next generation.
- SE - (μ + λ): μ in this case the parents are selected from the current generation and generate λ offspring through recombination operators and / or mutation where the best μ will be kept for the next generation.
- SE - (μ, λ) is the most used where μ parents will be selected from the current generation and used to generate λ children (with λ> = μ) and only the best individuals descendants of μ construct the next generation by destroying completely the parents.

2.4.5. The Procedure

The general structure of the SE algorithm - (μ, λ) is shown in the following pseudo-code

```
Evolutionary algorithm strategy SE -(μ, λ)
Mi: best individu ; NG{}: the list of children individus: CP:
cross pobability; MP: mutation probability; P: population;
Input: μ: the number of candidate solutions in the parent
generation.
                  λ: is the number of candidate solutions
generated from the parent generation
P                       generate-randomly (μ solutions potentiel)
```

```
For each individu Ni ∈ P do
      Evaluat(Ni)
End
Mi                return-best-individu(P)
while (no-stopping-cretirien) do
           NG{}                     ∅
    for i=0 à λ do
        PRi             return-parents (P, i)
       Si A<P_M alors NG {}      crossover (discret-coss,
intermediate-coss)
       If A<P_C then NG {}        mutation (mutation (strategy
parametes), mutation (décision-parametes))
    endfor
   Evaluation(NG)
   Mi            return-best-individu (NG, Mi)
   P              selection (P, NG, μ)
end
output: Mi
```

2.5. Differential Evolution (DE)

Another paradigm in the family of EA is the differential evolution (DE) proposed by Storn and Price in 1997 (Storn, 1997). The DE algorithm is similar to genetic algorithms since it is grounded on a population of individuals to research an optimal solution. The main difference is that in the GAs, the mutation is the result of small disturbances to the genes of an individual, but in the DE the mutation is the result of arithmetic combinations of individuals.

The DE is easy to implement that requires less adjustment parameters such as the weighting factor. It is considered as a reliable optimization technique, precise, robust and fast. Unfortunately, according to Crick & al in 2004, noise can interfere with the performance of DE because of its greedy nature of the DE and self-adaptation can eliminate the need for manual adjustment of control settings.

2.5.1. The General Process

The Differential Evolution algorithm involves maintaining a population of candidate solutions subjected to iterations of recombination, evaluation, and selection. The recombination approach involves the creation of new candidate solution components based on the weighted difference between two randomly selected population members added to a third population member. This perturbs population members relative to the spread of the broader population. In conjunction with the selection, the perturbation effect self-organizes the sampling of the problem space, bending it to known areas of interest.

2.5.2. The Procedure

The general structure of the DE algorithm is exhibited in the next pseudo-code:

```
Differential Evolution (DE) algorithm
Input: Population_size, P roblem_size, Weighting_factor, Crossover_rate
Population              InitializePopulation(Population_size,
Problem_size);
EvaluatePopulation(Population);
Sbest                GetBestSolution(Population);
while:not StopCondition() do
     NewPopulation             ∅
     For each Pi ∈ Population do
                   Si              NewSample(Pi, Population,
Problem_size, Weighting_factor, Crossover_rate);
                  if Cost(Si) ≤ Cost(Pi) then    NewPopulation
Si;
                         else   NewPopulation
Pi;
                  end
      end
      Population              NewPopulation;
      EvaluatePopulation(Population);
      S_best                  GetBestSolution(Population);
end
Return: S_best;
```

2.6. Paddy Field Algorithm (PFA)

It is a new algorithm proposed by Premaratne & al in 2009 (Premaratne, 2009), which works on a principle of reproduction depends on the proximity of the overall solution and a population density similar to the plant populations. Unlike evolutionary algorithms, it does not use a combined behaviour or cross between individuals, it uses the pollination and dispersal steps. The PFA consists of five basic steps (seed, selection, planting, pollination, dispersion). It has a lower cost of computing.

2.6.1. The Procedure

The general structure of the PFA is exhibited in the next pseudo-code:

```
Paddy Field Algorithm (PFA)
Input: population_size, Problem_size
Initial population              randomlly scattering seeds
(plants)
While not stopping cretirion do
     Selection                  select-best-plant(population)
    For each plant ∈population do
         Seeding (population)
         Pollination population)
     End
        Best_solution                evaluation population
         Population                     dispersion (population)
Return:best_solution
```

2.7. Analytic Comparison Between Evolutionary Algorithms

This section presents a comparative analysis of the evolutionary algorithms (EAs) seen so far in terms of the representation, operators, areas of application and control parameters.

3. Swarm Intelligence Algorithm (SIAs)

The swarm intelligence (SI) is a phrase introduced firstly by Gerardo Beni and Jing Wang in 1989, as part of cellular robotic systems (Beni, 1993). The SI is a new paradigm emerging in bio-inspired computing that allows the implementation of adaptive algorithms inspired by the collective behaviour of natural swarms as a tool for solving complex problems. It can be seen in groups of birds, schools of fish, as well as insects like mosquitoes and gnats. As has mentioned in the previous section the EAs are based on the genetic adaptation of organisms, but SIAs are based on social behaviour and collective intelligence of organisms.

Generally, An SIA is composed of a population of simple autonomous agents interacting locally between them and with their environment. Agents are subsystems

Table 1. The analytic comprison between evolutionary algorithms

	Steps	Parameters	Input and Coding	Application domains
Gas	- Initialisation of population. - Reproduction operators (selection, crossover (1-point, 2-points, uniform), and mutation (exchange)), - Replacement.	- Number of iteration. - Size of the population. - Crossover probability, mutation probability - Size of the chromosome.	- Individuals (potential solution) - Binary, real or matrix.	Information retrieval, classification, clustering, Traveling salesmen problem, Business intelligence, data mining problem, network planning, scheduling problem, image segmentation, electromagnetism and mechanics computational fluid.
GP	- Population of initial program, - Tournament selection. - Tree crossing. - Mutation (change under a tree of a program) - Replacement	- Number of iteration. - The population size. - Crossover probability. - Mutation probability	-Nested list (syntax tree with leaves or terminals, internal nodes called function).	Automatic discovery of reusable subroutines, classification, rules of associations, image processing, medical knowledge extraction, object detection, soccer robot problem
ES	- Mutation parameter - Selection (deterministic process). - Discrete crossover for strategic parameters. - Intermediate crossing for the decision parameters.	- Stopping criterion. - Probability of crossover. - Probability of mutation. - Selection and form of sampling used.	-Individuals (strategic parameter, decision parameter) -Vector with real components	Machine vision system, image processing, decision support system, scheduling problem, the problem of wireless mobile networks.
DS	- Selection. - Crossover. - Mutation (arithmetic combination between individuals). - Replace.	- The size of the population. - The size of the problem. - crossover probability.	-Vectors of real values.	Unsupervised image classification, clustering, digital filter design, optimization of nonlinear functions.
PFA	- The dissemination of seeds. - Planting, selection. - Seeding, dispersion.	- Number of initial seed. - Seed size - Selection threshold. - the initial value of number of - Seed	-Individuals (potential solutions) -Linear=[x1,x2..x]	Optimization of neural network parameters, adjustment parameters in the PID controller, the optimization of the continuous function.

that follow very simple local rules, and although there is no centralized control structure dictating how individual agents should behave. The local interaction between these agents led to the emergence of a complex global behaviour that achieves a goal. One type of self-organization emerges from the set of actions of the group. This class of algorithms can produce solutions that are fast, robust and with less cost to more complex problems.

Since 1990, several algorithms inspired collective behaviour (such as social insects, flocking birds) have been proposed in the literature, and applied successfully in many application in real life (like the traveling salesman problem, routing network, classification, data mining, scheduling tasks, and collective robotics ...... .etc. In this section we discuss in detail a set of (SIAs):

3.1. Particles Swarm Optimization (PSO) Algorithm

3.1.1. The Origin Of The Algorithm

The PSO is an SI stochastic algorithm that was proposed by the psychologist James Kennedy and the electrical engineer Russell Elberhart in 1995 (Kennedy, 1995). These two researchers used the studies of Craig Reynolds in the late 1987 about the foraging of a birds group (Reynolds, 1987). This algorithm was originally used for solving nonlinear optimization and continuous problems, but more recently it has been applied to many problems of real life, such as dynamic monitoring systems (Eberhart, 2001), control collective power (Yoshida, 2000), and the composition of music (Blackwell, 2002).

3.1.2. The Inspiration Source

The PSO is inspired by a sociological phenomenon associated to the flocking of birds and precisely the foraging behaviour. Naturally the birds fly in a group without collision and making use of their efforts to maintain an optimal distance between them and their neighbours as well as a local interaction to achieve the common objective of the group.

The phenomenon of flocking is clearly defined in the following scenario: Initially, a group of birds is looking randomly for food in a zone. There is only a piece of food in the desired area. The birds do not know where is the food, but they know how far away and the positions of their neighbours. So, what is the best strategy to find food? An effective strategy is to follow the bird that is the nearest food.

3.1.3. Summary of the Algorithm:

The PSO is a global optimization algorithm generally used to deal with problems where a better solution can be represented as a point or a surface in an N-dimensional space. It is based on a population of solutions (particles) random (each solution is like a bird) and each particle is associated with a velocity and position. As time goes, the particles move through the search space with dynamically adjusted speeds based on their performance and updating each solution to direct the research process. Therefore the particles tend to fly in an optimal search area where they are evaluated according to certain fitness criteria after each step in time and also through simple

Figure 4. The various steps of the operation of particle swarm optimization algorithm (PSOA)

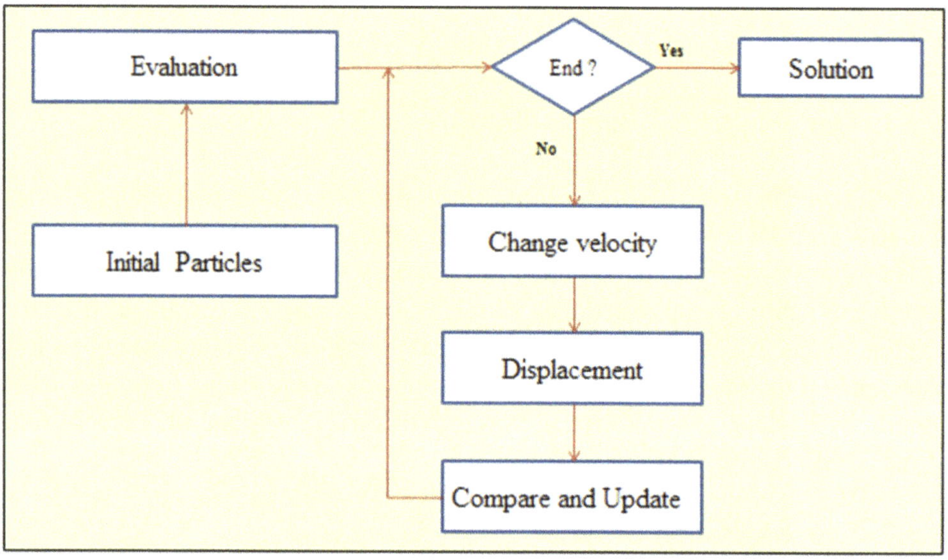

rules following the particles with the best solutions (best fitness value) to be gradually converging to a local minimum.

3.1.4. The General Process

Each particle is characterized by its position, its best position found, the best position funded by its neighbors and its velocity. During the research process, each particle adjusts its position in the search space basing on his best position finding (Pbest) and the best position found by his neighbors (g_{best}). The general process of PSO is presented in the following figure:

As shown in previous figure each step will be detailed in the next:

3.1.4.1. The Initial Population

The generating of N initial solutions is based on the random generation of the velocity and position of each particle.

3.1.4.2. Evaluation

This step evaluates the quality of each solution by evaluating the position of each particle to detect the best position.

3.1.4.3. Change Velocity

Iteratively, each particle changes its speed depending on its current position and the better position of its neighbourhood using the following equation:

$$V_{K+1}^{i} = V_{K}^{i} + C_1 r_1 (P_{K}^{i} - X_{K}^{i}) + C_2 r_2 (P_{K}^{g} - X_{K}^{i})$$

- P_{K}^{i} : The best position X_{K}^{i} of the particle i in the iteration K.
- P_{K}^{g} : The best position among all the positions of the particles.
- $C_{1,}$ C_2: Two learning factors that control the influence of social and cognitive components.
- $r_{1,}$ r_2: two random numbers generated independently in the range [0,1].

3.1.4.4. Displacement

The movement of each particle from one position to another is based on the current position and its speed of displacement. The next position of each particle is calculated through the following equation:

$$X_{K+1}^{i} = X_{K}^{i} + V_{K+1}^{i}$$

- X_{K}^{i} : The position of the particle i in the iteration K.
- V_{K}^{i} : Velocity of the particle i at iteration K.

3.1.4.5. Compare and Update

We compare the current position of each particle X_{K+1}^{i} with his best position found in the previous steps P_{K}^{i}. if $P_{K}^{i} \leq X_{K+1}^{i}$ then $P_{K+1}^{i} = X_{K+1}^{i}$.

3.1.4.6. Procedure

The general structure of the particle swarm optimization algorithm is presented in the following pseudo code:

```
Particles swarm optimisation
P best: best solution
Input:
-            N:the number of particles of the initial population
```

```
-         The fitness function
P             generate-randomly (N, Vi, Xi)
For each Pi do
       P^g_0                  Evaluat (pi)
end
P^i_0                   X^i_0
K             0
While (no stopping cetirien) do
       For each particle pi of the population do
                  Change-velocity (pi)
                  Displacement (pi)
            P^i_{K+1}                Comparer ( P^i_K, X^i_{K+1})
                 Mise à jour ( P^i_{K+1})
          end
                    P^g_{K+1}                Evaluat(Pi)
                              P_best                  P^g_{K+1}
                                   K              k+1
    end
output: P_best
```

3.2. Ant Colony Optimization Algorithm

3.2.1. The Origin of the Algorithm

The ACO algorithm is a meta-heuristic probabilistic SI which was introduced by an Italian Marco Dorigo around the 1990 during his doctoral thesis on the optimization and natural algorithms (Dorigo, 1992). This algorithm was firstly proposed for solving a discrete optimization problem at the end of 1991, and also used for solving the traveling salesman problem in (Dorigo, 1997). Several versions of this algorithm have been proposed in the literature as the minimisation algorithm proposed by -maximum Stützle and Hoos in 1997 (Stützle, 1997), and the ant system algorithm proposed in (Dorigo, 1996).

3.2.2. The Inspiration Source

The ACO was inspired by the behavior of ant colonies that are social insects with the ability to find the shortest path between their nest and the food source as well as the phenomenon of stegmergy introduced in 1959 by fat in (Grasse, 1967) which refers to the indirect communication between ants using pheromone trails that allows them to find the most current paths.

Ants first start by randomly explore the environment around their nest. When moving, ants leave a chemical pheromone trail on their way, the movement of ants is guided by the smell of pheromone Ants first start by randomly explore the environment around their nest. The movement of ants is guided by the smell of pheromone. Ants tend to choose the ways marked by the concentration of the stronger pheromone. In return, the amount of pheromone that leaves the ant on its way depends on the quality and quantity of food wear.

3.2.3. Summary of the Algorithm

The main idea of the algorithm is to model the problem to be solved as a research problem of an optimal path in a weighted graph by using artificial ants to find the quality paths among all the paths that exist. The ants move from one node to the next following a transition rule based on a probability of displacement. Each solution built is evaluated according to a fitness function in order to detect the shortest path (the nodes of the graph which corresponds to the components of the solution).

3.2.4. General Process

The general operation of ACO is presented in the following figure and each step is detailed in the next:

3.2.4.1. Coding Solutions

The solutions space of the problem is represented as a graph where each solution is a path.

3.2.4.2. Creation of Heuristic Solutions

The creation of N initial solutions using heuristics based on external resources. This step will generate values for the arcs that connect the nodes.

3.2.4.3. Pheromone Initialization

Initially, the traces of all tracks of the graph pheromone are initialized with a small constant value.

Figure 5. The functioning of the ant colony algorithm (ACO)

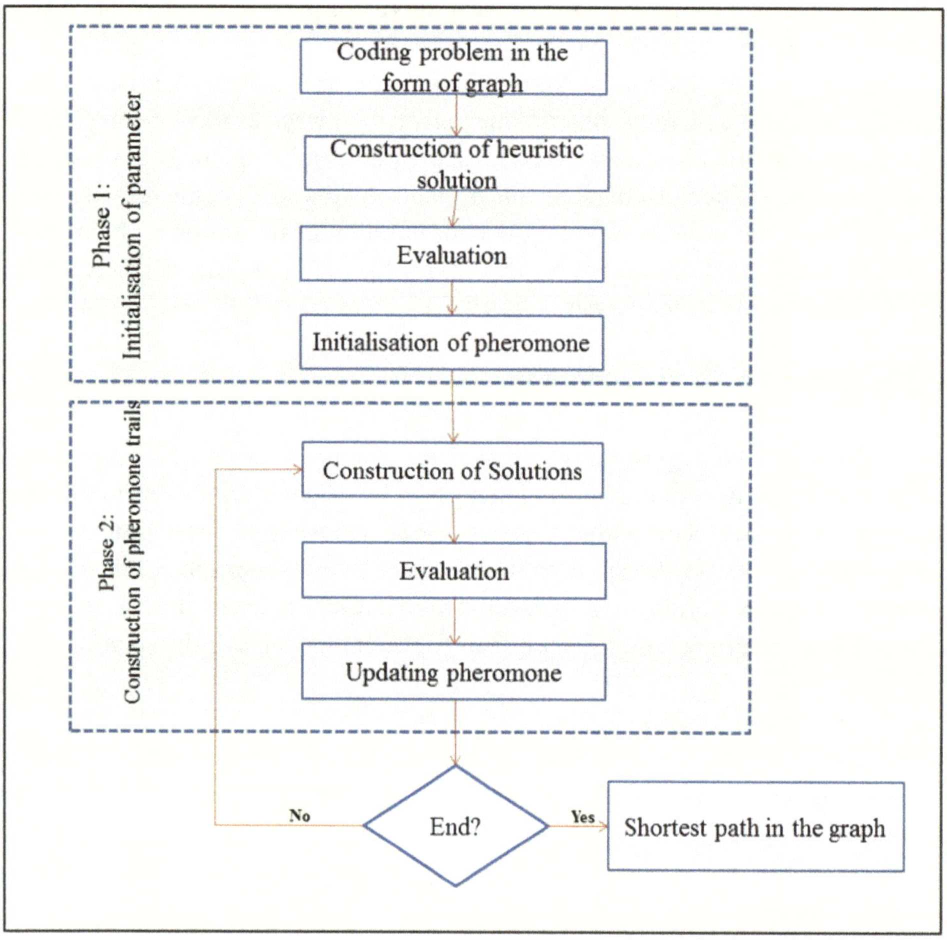

3.2.4.4. Solutions Construction

Artificial ants move iteratively through neighbouring nodes of the graph according to a transition rule. This movement is carried with a probability rule, which allows for each ant to sequentially select components (nodes) of the solution using the pheromone intensity of each track. The solution of each ant is built when all components of the solution are chosen by this ant (for example, when the ant has completed a tour / full path on the graph of the problem). The probability of movement of each ant from the node i to the node j depends on the following equation:

$$P_{ij}^{K}(t) = \frac{\left[\tau_{ij}(t)\right]^{\alpha} * \left[\mu_{ij}\right]^{\beta}}{\sum_{l \in N_i^k} \left[\tau_{il}(t)\right]^{\alpha} * \left[\mu_{il}\right]^{\beta}}$$

- P_{ij}^{K} : is the probability of the kth ant to move from node i to node j in t[th] iteration.
- N_{i}^{k} : is the set of nodes in the neighbourhood of the Nth ant in the ith node.
- $\tau_{ij}(t)]^{\alpha}$: is the amount of pheromone on the arc connecting the node i and node j, weighted by α (the confession of the heuristic).
- : is the heuristic value of the arc connecting the node i and node j, weighted by β. This value does not change during the movement of the ant and during the execution of the algorithm.
- : is the heuristic information, or a path of visibility, which is a priori information about the problem provided by different source other than ants.

et are weights parameters that control the relative importance of pheromone against heuristic information.

3.2.4.5. Evaluation

Once a solution (path) is built, it will be evaluated using a fitness function to determine the amount of pheromone deposited on each track.

3.2.4.6. Update Pheromone

Once complete solutions have been built, the pheromones slopes must be updated. This step also includes the evaporation of pheromone (the pheromone potentials are reduced) in order to help the ants to forget the bad solutions that were previously learned. The values of the pheromone of the visited arcs are increased, with quantities proportionally inverse to the cost of their tours.

(1-P)*(t)+

, if arc(i,j))

=

else 0

- Q: is a specific constant.
- M: is the number of ants.
- A: is the set of arcs of the graph.
- P: the evaporation rate of pheromone.

- : is the function of the overall cost of the tour) built by the kth ant at iteration t..
-): is the set of arcs visited by ant k at iteration t.

3.2.5. Procedure

The following pseudo code summarizes the general structure and the different stages of the ACO algorithm

```
Ant Colony optimisation procedure
SH: heuristic solution.
Input: N: the size of the population, m, P,
Coding problem (representation of solutions in the form of
weighting graph)
Sh                        creation of heuristic solution
F(cbest)                          Evaluate (Sh)
cbest                              Sh
Pheromone                   initialisation of pheromone
While (no stopping criterion) do
       S
      For i=1 to m do
                      Si              probability (pheromone,)
                    F(si)                        evaluate(Si)
             If (f(Si) f(cbest)) then
                   f(cbest)                f(Si)
                   cbest                  Si
             end if
                S                    Si
      End
      Evaporation-of-pheromone(pheromone, P)
      For each Si  S do
               Updating-pheromone(pheromone, Si,f(Si))
       end
Output: cbest
```

3.3. Bacterial Foraging Optimization (BFO)

3.3.1. The Origin of the Algorithm

The bacteria are characterized by diverse behaviours like (foraging, reproduction and movement) that summers the inspiration of many computer scientists for a long time for the development of optimization algorithms. In 2012 an algorithm called bacterial foraging optimization (BFO) was introduced by Passino & al in (Pasino, 2012) where they are used studies conducted by biologists on the chemotactic movement of E.coli bacteria. This algorithm represents a promising new branch in bio mimicry that can bridge the gap between microbiology and engineering.

3.3.2. The Inspiration

The BFO is the first algorithm based on the foraging and moving of a bacteria colony (such as E. coli). The bacteria are always looking for the nutrients to maximize the energy obtained per unit of time and communicate by sending signals. The movement of bacteria is called chemotaxis, which represents the key idea of this algorithm.

3.3.3. The Summary of the Algorithm

The input of the BFO is a set of artificial bacteria that will be displaced in the solution space simultaneously taking into account their performance and their proximity with others bacteria. Following successive stages (moving, swarming, reproduction, disposal and dispersal) the population of bacteria searches the best solution to a problem from iteration to another until it reaches the stop criterion.

3.3.4. The General Process

The basic functioning of BFO is shown in the following figure:

The figure above shows the general functioning of BFO and the role of each step in this process are detailed below:

3.3.4.1. Initialization of Locations

Initially the location of each bacterium must be initialized randomly based on the number of artificial bacteria and the problem to be optimized.

Figure 6. The general process of the bacterial foraging optimisation algorithm (AOBB)

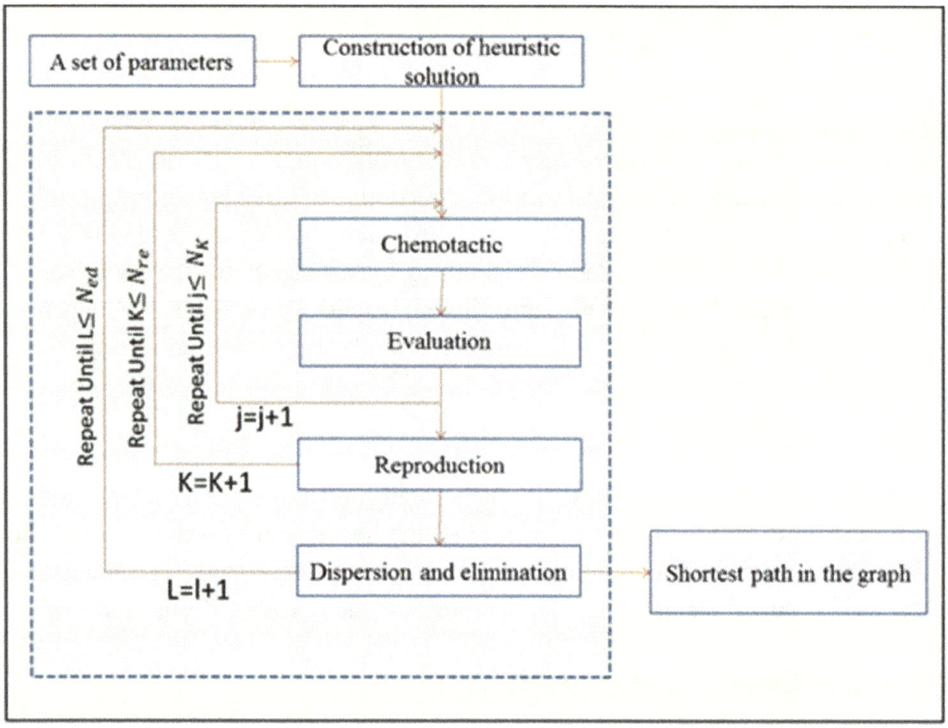

3.3.4.2. Chemotaxis

The movement of E. Coli bacteria is guided by two characteristics: they are unable to choose the direction in which they move and they are also unable to follow the same direction a long time because they easily forget the directions in which they are going. For this reason there are two different types of motions:

- **The oriented movement (swimming):**

It is the movement of bacteria in the same direction. The presence of a chemical guardian (place of food) allows bacteria to react with the phenomenon of chemotactic and they will direct their movements in the direction of such guardian. In the case where a bacterium feels that it moves in the right direction (towards the chemical caretaker) then it continues straight as for a long time before tumbling phase.

- **The random motion (tumbling)**

It is the movement in a random direction where if a bacterium forgets the direction in which it goes and if she feels that she is not going in the right direction, it will follow a new direction chosen randomly. The movement of the i^{em} bacterium after each step is given by the following two equations (Passino 2002):

=+C(i)

- Is the location of the i^{em} bacterium in the P^{eme} chemotactic, K^{em} reproduction, the removal and dispersion step.
- C(i) the length of the walking unit.
- the direction angle of the Pem step.

3.3.4.3. Swarming

Each bacterium sends attractive signals to other bacteria in order to be grouped and send also repellent products to signal others to be at a minimum distance from her. The position (cell) signaling can be represented by the following equation (Passino 2002):

J_{cc}(=

+

- J_{cc}(: the objective function
- S: the total number of bacteria (number of cell in a population)
- P: the number of variables to be optimized which are present in each bacterium
- =: refers to a point in the dimension of the search area
- : the depth of the bait released by the cell
- : a measure of the width of the attraction signal
- : the height of the repellent effect
- the measurement of the width of repulsive

3.3.4.4. Reproduction

After the NC steps of chemotactic, the reproducing step must be done and the performances (fitness function values) of bacteria will be stored in an ascending order. The functioning principle is that the bacteria with less healthy eventually die and the remaining bacteria (i.e. healthy bacteria) will be divided into two identical and placed in the same place.

3.3.4.5. Dispersion and Elimination

In order to avoid local optima, a process of dispersion and removal is done after the REN phases of reproduction. Depending on a probability (ped), a bacterium is selected to be scattered and moved to another position within the environment.

3.3.5. Procedure

The general structure of BFO is grouped in the following pseudo code:

```
Bacterial foraging optimisation (BFO)
Num the size of the population ;  Ned: the number of elimination
and dispertion step; REN: the number of reproduction step ; Nc
is the number of steps of chemotactic; Ped: the probability of
a cell being subjected to the removal and dispersion.
Input:
Num, Ned, REN, Nc, Ped,
P                  initialize the location of each bacterium
For l=0 to Ned do
     for K=0 to REN do
           for j=0 to NC do
                     P              chemotaxis (p)
                  For i=1 to Num do
                       Calculate the fitness function for each
bacterium f(i)
                            If   Bbest  f(i) then   Bbest
f(i)
                  end
          end
               P                    reproduction(P)
      end
        for each bacterium i from the popualtion P do
                    If A Ped then
                    P(i)                    Dispersion and
elimination (I)
           end
end
output: Bbest
```

3.3.6. Other Bacteria Algorithms

In the literature others algorithms have been proposed by mimicking the biological characteristics of bacteria such as:

- The algorithm of bacterial chimiotaxis (BCC) proposed by Li & al in 2005 (Li, 2005).
- The algorithm Superbag Algorithm (SUA) has been proposed by (Anandarman, 2012) which is based on the findings of the epidemic and antiviral research. The operation of the SUA is comprised from the following operators (initialization of the population, inversion mutation with inter changes for each pair of bacteria, transfer genes between bacteria to improve the level of fitness).
- The Bacterial Optimization Algorithm (BOA) was proposed by Niu and Wang in 2012 (Niu, 2012). The functioning of the BOA consists of the following steps: chemotaxis, communication, elimination, reproduction and migration.

3.4. Wolf Colony Algorithm (WCA)

3.4.1. The Origin of the Algorithm

The WCA was firstly introduced by Liu & al in 2011 (Liu, 2011) which is based on the studies conducted by Macdonald in (Macdonald, 2004) and Fuller in (Fuller, 2003).

3.4.2. The Inspiration Source

The wolf colony algorithm (WCA) is inspired by the natural social life of wolves and intra-specific phenomenon of social hunting. It mimics the operation from intelligent predatory of the northern gray wolves as (seeking food, besieging behaviour and the behaviour of the division food).

3.4.3. The Summary of the Algorithm

In the WCA each artificial wolf is a solution. Initially a set of artificial wolves are generated for searching in the field activity of a prey flock (the solutions search space). When the artificial search wolves discovered that herd, they inform other artificial wolves of this position by the howling (the solutions will be improved iteratively). Other artificial wolves get closer to that flock and besiege. The wolf colony is updated iteratively. The performance of the WCA is discussed based on the objective of the optimization problem.

Figure 7. The functioning of the wolf colony algorithm (WCA)

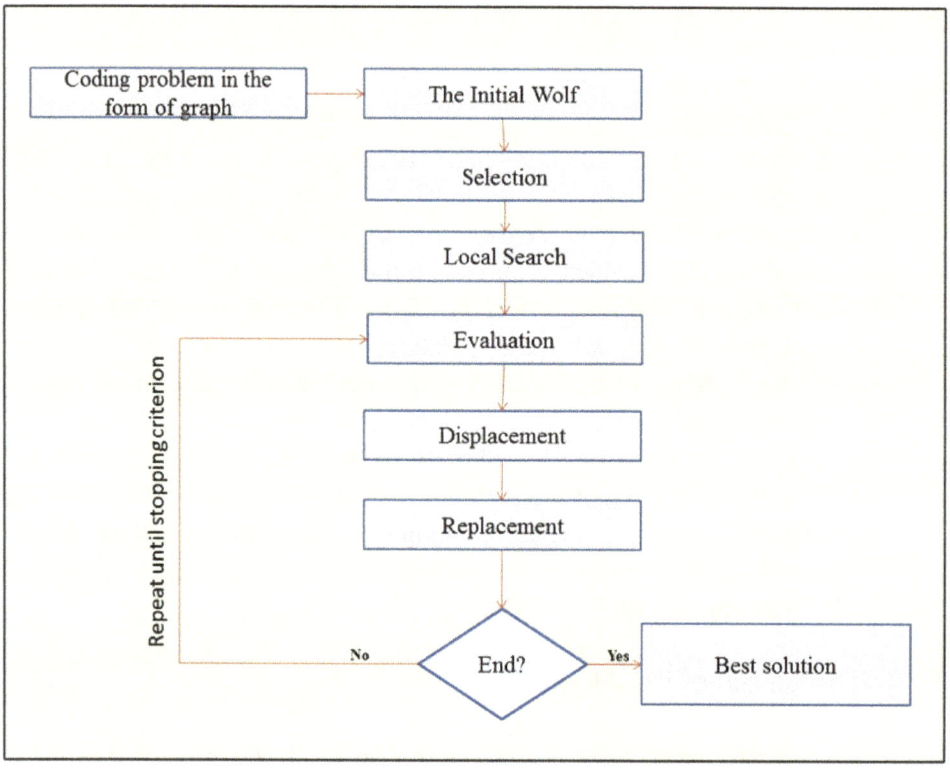

3.4.4. General Process

The operation of the WCA is shown in the following figure:

The figure above describes the relationship between the different steps involved in the operation of the WCA, which will be detailed in the next:

3.4.4.1. Initialisation of Wolf Colony

D is the size of the search space, n is the number of individual; Xi represents the position of artificial i^{eme} wolf, so we have the following equation to represent the position vector of each wolf (Liu, 2011):

$$X_i=X_{i1},.......X_{id},......X_D \quad 1.$$

with 1i et 1dD

This step allows the initializes of N artificial wolves by randomly generating the position Xi of each one of them.

3.4.4.2. Selection

This operator selects from N wolves, q most optimum artificial wolves (called Scouts wolves that are closer to the prey flock) designated for research to detect the potential activities of the prey flock in order to increase the probability of discovering their positions.

3.4.4.3. The Local Search (Besiege The Flock Of Prey)

Each search wolf selected is moved using the following equation:

$Yj = XX_i + randn .stepa$

- Maxdh: denotes the number of maximum research
- XXI: Is the location of the ith Scout wolf (totally h locations are created around the candidate wolf).
- Yj: the j^{eme} position of research.
- Rand: A random number in the range [-1, 1].
- Stepa: the search step

This step is completed when the number of searches is greater than maxth where the current location of the search is better than the optimal search location fix in advance.

3.4.4.4. Evaluation

This step is for evaluating the performances and the positions of wolves in order to compare and detect the best position using the objective function.

3.4.4.5. The Displacement of Artificial Wolves (The behaviour of besieging)

When the prey flock is found, the position of each wolf will be updated and the howling is used to notify other wolves about this position. The displacement of each wolf is produced according to the following equation:

=+rand.stepb.(

- : is the position of the prey flock after the iteration K and in the dme dimension of the search space.
- : the position of the i^{eme} artificial wolf
- Rand: a random number in the range [0, 1].
- Stepb: the search step.
- K: the iteration number.
- The range [$XMIN_d$, $XMAX_d$] is used for the representation of the dth position.

3.4.4.6. Replacement (Division of Food)

The food is eaten first by the strongest wolves, then by other wolves. In this colony, the m (parameter set in advance) wolves with the worst performance will be replaced with new artificial wolves created randomly and will be integrated into the next iteration of the colony. This mechanism can prevent the algorithm falling into local optimum.

3.4.5. Procedure

The general structure of the wolf colony algorithm (ACL) and the relationship between these function is present in the pseudocode:

```
Wolf colony algorithm (WCA)
Lbest: best solution ; CL: wolf colony ; LR{}: the list of
search wolf generated.
Input:
N, q, h:the direction of research, maxdh, stepa, stepb, m: the
number of wolves having the best performances which will be
kept for the next iteration.
CL                          initialisation (N, Xi)
For each wolf Li do
            Lbest                            evaluation (Xi)
While (no-stopping cretirien) do
           For each wolf Li  CL do
                LR{}                    Select(q)
          end
          for each wolf Li LR{} do
            Displacement following:  Yj = XXi + randn .stepa
            Li                      new position
          end
        Lbest                           evaluation(Xi)
```

```
        For each Li  CL
            displace (Li) using
           =+rand.stepb.(
        end
        Replacement (m)
end
output: Lbest
```

3.5. Wolf Packs Search (WPS)

There are other algorithms inspired by the social life of wolves such as wolf pack search (WPS) which is introduced in 2007 by Yang in (Yang, 2007). The WPS is composed of 4 main steps: initialization parameters ii) random generation of a set of packs of artificial wolf ii) evaluating the performance of each artificial wolf to determine the best solution using a a fitness function iiii) the circulation and update step.

3.6. Bat algorithm (BA)

3.6.1. The Origin of the Algorithm

The bat algorithm (BA) was proposed by Xin-She Yang in 2010 (Yang, 2010). It was firstly applied on the benchmark functions, and it gets better results compared to genetic algorithms and PSO.

3.6.2. The Source Inspiration

The bats have mysterious behaviours that have a long time ago attracted the attention of human. They emit a very loud noise pulse and listen to the echo that bounces from surrounding objects. Their impulses vary according to their hunting strategies. This behaviour is called echolocation, which allows to bats to find their prey and discriminate different types of insects even in complete darkness. The inspiration of the BA is summarized in the Bat movement for foraging and the behaviour of echolocation.

3.6.3. The Summary of the Algorithm

The input of the BA is a population of N bats (possible solutions) that uses echolocation to detect the distance. The artificial bats fly through a random search

Figure 8. The general functioning of the bat algorithm (BA)

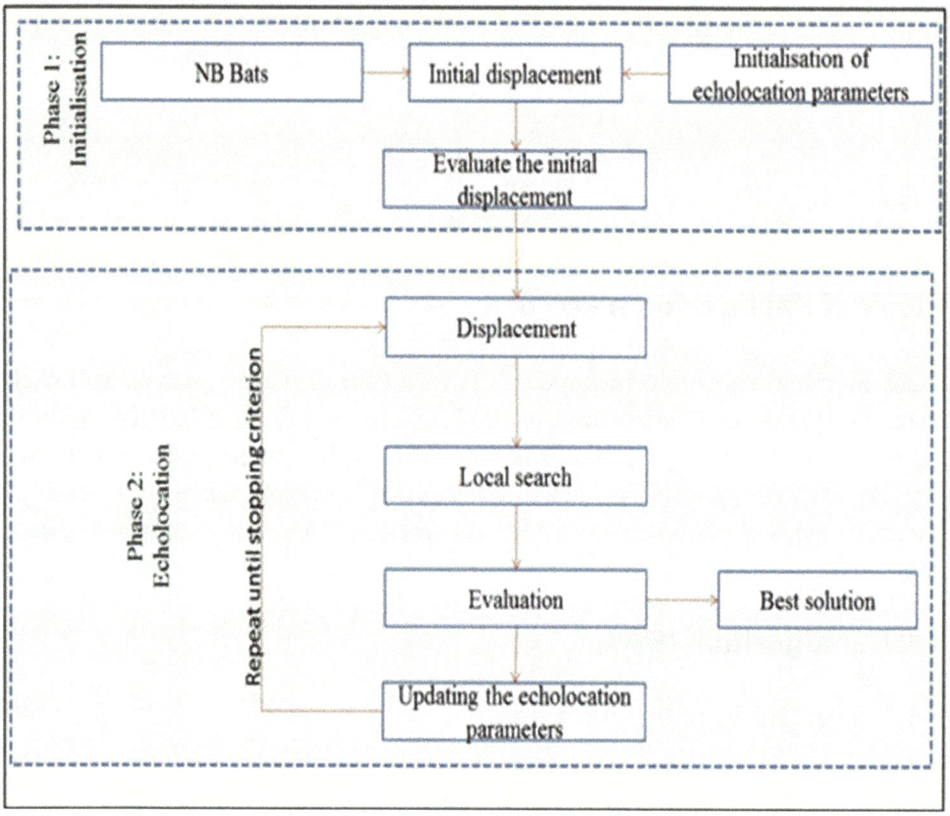

space by conducting the update of their positions Xi and velocities vi from one step to another. Each solution is evaluated by a fitness function. The goal is to find food / prey (best solutions).

3.6.4. The General Process

The operation of the BA is shown in the following figure. Initially, some numerical parameters associated with bats foraging behaviour are defined such as the velocity (vi) at the position (xi) with a fixed minimum frequency (f_{min}), a wavelength variable (λ) and intensity (Ao).

The previous figure demonstrated the various steps of the BA that will be detailed in the next:

3.6.4.1. Initial Displacement

Initially a population of N bats (potential solutions) must be initialized by the random generation of the positions and velocities of each bat.

3.6.4.2. Displacement

After the random displacement of bats in the previous step, the displacement of bats in the next step will be realized through the adjustment of their frequencies and updating of their velocities and positions. New locations / solutions (xit) and velocities (Vit) over time and step t are given by the following equation.

$$F_i = f_{min} + (f_{max} - f_{min}).,$$

$$V_i^t = V_i^{t-1} + (x_i^t -).$$

$$x_i^t = x_i^{t-1} + V_i^t$$

- [0, 1]: is a random vector drawn from the uniform distribution
- F_i: is the speed increment
- F_{min}, f_{max}: are the lower and upper bounds imposed for the frequency range of virtual bats.
- x_*: the best global position.

3.6.4.3. The Local Search

In this local search step a new solution is generated for each artificial bat of the current population according to the next equation

$$X_{nv} = X_{anc} + A^t$$

- : a random number.
- A^t: is the average intensity of all bats.

3.6.4.4. Evaluation

The evaluation of each artificial bat using the fitness function and the solutions were ranke in order to find the best solution X *.

3.6.4.5. Updating the Echolocation Parameters

The general rule is that the intensity decreases as the pulse emission rate increases, when a bat found his prey. The intensity and the pulsing rate are updated through the following equation:

$$A_i^{t+1}=A_i^t$$

$$R_i^{t+1}=r_i^0[1\text{-}exp(\text{-}t)]$$

- A_i^t, A_i^{t+1}: the previous and current intensity of the ith bat.
- et: constants with 0<<1 and >0.
- R_i^{t+1}: is the pulse rate of the i[me] bat at the iteration t+1.

3.6.5. The Procedure

The general structure of the bat algorithm is shown in the following pseudo-code:

```
Bat algorithm (BA)
PCS: population of artificiel bats
Entré: Ai, Fi, Ri, NB: the number of generated bats
*Initial displacement*
PCS                     initialisation of the population(Xi,
Vi, NB)
Evaluat(PCS)
While (no-stopping criterien) do
              PCS                 displacement(PCS)
                PCS             local-seach(PCS)
            Evaluate (PCS)
            updating echolocation parameters (Ai, Ri, fi)
            X*               Evaluat(PCS)
End
Output: X*
```

Figure 9. Description of the experience

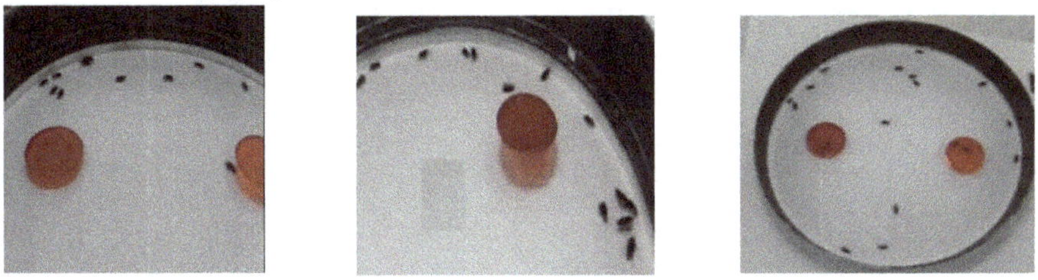

a- two Shelters b-exploration phase c- the experience environment

3.7. Cockroaches Colony Algorithm (CCA)

3.7.1. The Origin of the Algorithm

The CCA was introduced the first time by Bouarara & al in 2015 (Bouarara, 2015) where they have used the studies carried out by Bell & al on social life and behaviour of cockroaches proposed in (Bell, 2007).

3.7.2. The Inspiration Source

The CCA is inspired by the natural behaviour of cockroaches and the phenomenon of seeking the most attractive and secure place (shelter) for hiding. We can identify different types of cockroaches in our work, we are interested by the cockroaches that live in apartments, which are fertile and they are never isolated.

This phenomenon is well detailed in an experiment conducted by French biologists when they met a group of cockroaches in a basin where there's light everywhere, and they built two artificial shelters (shelter is a place with less brightness as shown in the figure) using two red circles because cockroaches do not observe the color red as shown in the following figure.

From previous experience it was observed that cockroaches have a choice of two shelters to hide where they always choose the most secure shelter. A biological model explaining this phenomenon is presented by the following:

3.7.2.1. Random Displacement of Cockroaches

Initially cockroaches will move randomly in all directions (exploration phase) as demonstrated in Figure 5.b. When a cockroach finds an attractive shelter, it hides and sends pheromones as smell to its congeners. The movement of cockroaches is guided by a set of displacement rules:

Figure 10. The groping of cockroaches under the same place (Bouarara, 2015)

a) Global view b) Detailed view

3.7.2.2. The Darkness Shelters

Cockroaches are attracted by the darkest places like corners and shelters with less brightness. The degree of darkness plays a very important role for the quality security of each shelter.

3.7.2.3. The Congener's Attraction

Each cockroach seeks shelter where there are more of its congeners (cockroaches from the same colony) to hide it.

3.7.2.4. The Security Quality

Cockroaches positioned in the middle of the shelter have a higher safety compared with cockroaches positioned at the border of the shelter.

3.7.3. Summary of the Algorithm:

Initially a set of cockroaches will be assigned randomly to N shelters (initial solutions). Iteratively these solutions going to be improved using the displacement process following a set of rules (the darkness of each shelter, the attraction of congeners, permutation, and quality safety) until a stopping criterion. The main objective is that every cockroach should be hidden in the shelter where he feels secure (find cockroaches position which ensures the best solution to a given problem).

Figure 11. The general functioning of cockroaches colony algorithm (CCA)

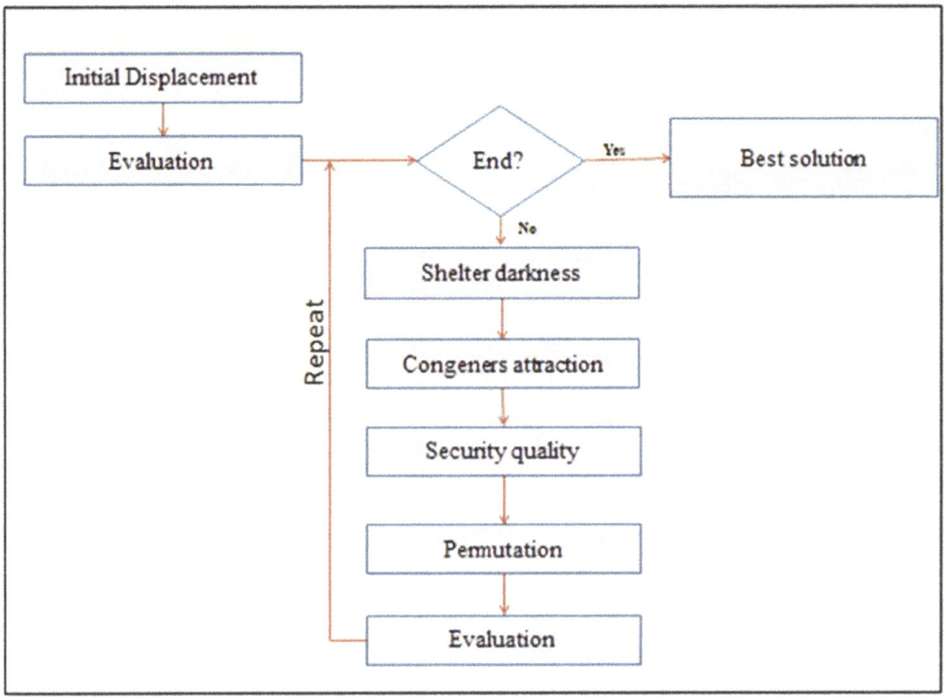

3.7.4. General Processes

3.7.4.1. Representation of the Problem

Initially the problem must be transformed as a set of solutions and each solution is a position (shelter) where cockroaches is hidden.

3.7.4.2. Initial Displacement

The initial displacement of artificial cockroaches is carried according to a random generation of their positions (shelter Xi where each cockroach was hidden).

3.7.4.3. Seeking Operators

After the random displacement initially performed, cockroaches begin to change their current shelter according to the following rules:

- **Shelter Darkness:**

The darkness of each shelter Xi for the cockroach Ck is calculated following the next equation:

$O_t(X_i)=O_t+$

- **N: the Cockroaches Number**
- : the number of cockroaches in the shelter Xi in the iteration T
- O_t (X_i): the darkness rate of the shelter Xi in the iteration T.
- **The Attraction Congeners:**

In each iteration, the attraction of the neighbourhood (congeners) for each cockroach in each shelter changes depending on its positions and the best position of its neighbourhood.

$+V_1(-) + V_2(M_T-)$

- : the best security quality of the cockroach K among all the shelters where it was hidden previously.
- : the rate of attraction of the neighbourhood (congeners) for the cockroach K hiding in the shelter Xi at iteration T.
- : the secuity cuality of the cockoach K in the iteration T.
- M_T: the best quality security among all security quality of cockroaches in all positions at time T.
- $V_{1,}$ V_2: [0,1].
- **The Security Quality:**

The security quality of each cockroach K in shelter Xi is calculated through the following equation:

$=+*O_{t+1}(X_i) +$

3.7.4.4. Evaluation

This step evaluates the position of each cockroach and the content of each shelter following an objective function fixed in advance in order to detect the best solution.

3.7.4.5. Permutation of Shelters

This step is used to avoid the local optimum and maintain the diversity of results. Firstly, a permutation probability (PP) is fixed and a random variable B is generated if B <M MP then cockroaches are randomly selected (one per shelter) and we permute them.

3.7.4.6. Displacement

Iteratively, each cockroach leaves his shelter to look for another place with more security. The displacement of cockroaches is achieved following a transition rule that allows for each cockroach to choose the direction and the next shelter that it will join. This choice is made via the following equation using the amount of security and external information that must be fixed in advance (e.g. in the traveling salesman problem that information is 1 / distance).

- Is the probability of the kth cockroach to pass from the shelter i to the shelter j in the iteration t.
- : is the set of shelters in the neighbourhood of the Kth cockroach in the ith shelter.
- : The security quality of the Kth if it moves to the shelter j at iteration t.
- : is the value of external information of the path that connects the shelter i and shelter j. This value does not change during the movement of the cockroach and during the execution of the algorithm.

3.7.4.7. Updating of Shelter

For each cockroach, we compare the security quality of its current position with its best position found in the previous steps. If the security quality of its current shelter is better than the security quality of its shelters previously visited, its position will be defined as the best position =.

3.7.5. Procedure

The next pseudo code summarizes the general structure and the different steps of the CCA algorithm:

```
Cockroaches colony algorithm (CCA)
Coding the problem
P                           generate-randomly (
M_position                           evaluation (P)
T                    0
```

```
While (no stopping criterion) do
          P            Ot+1(Xi)
           P
           P             (Xi)
            if A≤PP then
              choose randomly two cockroaches from two different
shelters
                P                   permute (,)
            End if
            M_position                          evaluation (P)
           For each cockoach of the population do
                     (T)
           End
            P                      displacement ()
             P                      update ()
               T                   T+1
   End
output: M_position
```

3.8. Mosquito Optimization Algorithm (MPA)

In (Feng, 2009), the traveling salesman problem is solved by the Mosquito Host Seeking Algorithm (MHSA) derived from the mosquito hunting behaviour.

3.9. Wasp Swarm Optimization (WSO)

In 2008 Rinkler & al imitated the behaviour of wasps for the development of a new algorithm called Wasp Swarm Optimization proposed in (Runkler, 2008).

3.10. Social Spiders' Algorithm (SSA)

Social spiders also served as inspiration for example for the detection in images (Bourjot 2003) or securing wireless networks (Benahmed, 2012).

3.11. The Cuckoo Search Algorithm (CSA)

The behaviour of social animals such as cuckoos (birds of the family Cuculidae) has also been imitated by researchers. The cuckoos have as particularity to parasitize the nests of other species by laying their eggs. Other birds have two options: they perceive the trickery and destroy eggs, or they did not perceive them. To ensure the survival of their offspring, cuckoos are seeking to place them in the nests where they have the greatest chance of survival. Gandomi & al have developed The Cuckoo Search algorithm that was directly inspired by this parasitism in (Gandomi, 2013).

3.12. Fireworks Algorithm (FWA)

Since its creation, the FWA has been developed, and many extensions have seen the light. It was proposed for the first time in 2010 by Tan and Zhu in (Tan, 2010). This first algorithm was designed for function optimization which consists of local search space. However, in 2011 the FWA algorithm was used by Janecek and Tan in (Janecek, 2011) for the improvement of multiplicative update of non-negative matrix factorization algorithm. They put the FWA in confrontation with four other approaches using nature inspired algorithms (Genetic algorithms, Particle Swarm Optimization, Differential Evolution, and Fish School Search). Meanwhile, the FWA was studied many times. In (Pei, 2012), the authors presented an empirical study about the effect of approximation approaches for the enhancement of FWA. The first improvement that the algorithm has seen was in 2013 when an improvement was presented in (Liu, 2013) in which authors presented a new improvement of FWA. The improvement touched the way that the algorithm uses to compute the number of sparks and fireworks by designing a transfer function in order to control the exploration and exploitation of fireworks.

In (Zheng, 2013), an enhanced firework algorithm was proposed. The enhancement was designed to cover up some of the disadvantages of the original algorithm on shiftig functions, and the high computational cost per iteration. They proved five adaptations: a minimal explosion check, a new operator for generation of sparks, a new mapping way that could hold the problem of sparks that are out of the space, a new operator for the generation of Gaussian sparks, and finally, a new operator for selections of locations.

Later, in 2014, two extensions of firework algorithm were published in the CEC' 2014. One of them is a dynamic firework algorithm that was presented in (Zheng, 2014). They built a model in order to deal with the limitations of the previous enhanced firework algorithm (EFWA) by integrating a dynamic explosion mechanism for the fireworks. Also, they proved that the computational efficiency of EFWA could be improved by removing one of its operators. The other is presented (Li, 2014), in which the amplitude computation is changed by an adaptive one in purpose of dealing with the lack of adaptability of the EFWA.

3.12.1. The Procedure

The next pseudo code summarises the general structure and the different steps of the FWA:

```
Fireworks algorithm (FWA)
Input: Randomly select N locations for fireworks
 while not stopping cretiriendo
       Set off N fireworks respectively at the N locations:
        for all fireworks xi do
             Calculate the number of sparks as Si
             Calculate the amplitude of sparks as Ai
         end
          //m is the number of sparks generated
       for k =1 to m do
                Randomly select a firework xi and generate a
spark
       end
       select the best spark and the other sparks according to
selection strategy
 end
Return: best spark
```

3.13. Analytic Comparison Between Swarm Intelligence Algorithms

The next table 2 summarises a comparative analysis of the swarm intelligence algorithms (SIAs) seen so far in terms of the representation, operators, areas of application and control parameters.

4. CONCLUSION

The bio-inspired methods, or more generally meta-heuristics, will be a new revolution in computer science. The scope of this field is really wide, which open a new area in the next generation of computing, modelling and engineering of the algorithm. Unfortunately, we must keep in mind that the imitation of a biological mechanism does not necessarily provide an advantage, especially when the technology cannot match the biology, or because the desired objective may be different from that of

Table 2. Summary of a comparative study between the swarm intelligence algorithms

	steps	Parameters	Input/coding	Application domains
PSO	- Particle initialization (speed and position of each particle) - Change in velocity - Particle displacement - Evaluation - Compare and Update	- Best position of each particle - Best position in the neighbourhood - Two random numbers generated independently - Two learning factors	-The positions and velocities of the particles. -Binary coding, Real, speed	Clustering, Classification in the multi-class databases, feature selection, composition of Web services, extraction of fuzzy neural network rules, the integration of planning and scheduling
ACO	- Initialization of parameters, - Coding the problem as a graph, Construction of heuristic solutions, Evaluation, - Initialisation of pheromone - Updating and evaporation of pheromone	- displacement probability of each ant at each node - The weights of the arcs that connect the nodes. - Evaporation rate	-Graph weighted binary -Matrice	Traveling salesman problem, problem backpack, vehicle routing problem, detecting intrusion, problems of knowledge extraction, coloration of graphs, project scheduling, water distribution systems design.
BFO	- Initialise the positions of bacteria. - chemotaxis - The oriented movement - Random motion - Swarming. - Reproduction. - Dispersion.	- Number of bacteria. - The length of the walk unit. - Size of the search space. - Number of steps of chemotactic. - Number of disposals and dispersion steps. - Number of reproductive steps. - Probability of elimination.	- A set of i artificial bacteria	Extracting fuzzy rules, traveling salesman problem, classification, information retrieval, image segmentation, the resolution of the dynamic economic problem of distribution, applications on multi-objective optimization.
WCA	- Initialisation the positions of wolves. - Selecting scout wolf. - Local Search. - Evaluation. - The behaviour of besieging (displacement of the wolves).	- The number of scouts' wolves selected. - h: the search direction, maxdh, stepa, STEPB, - m: number of wolves that have better performance. - N: number of artificial wolves	-A set of artificial wolfs. -Integer, real, binary and matrix	The global optimization problems without constraints, the knowledge extraction tasks, Vehicle routing problems, combinatorial optimization problems, the detection of anomaly, the image segmentation.
FWA	- Selection of N locations - Explosions of N fireworks - Get the location of sparks - Evaluating the position of the sparks - Calculate the amplitude - Mapping - selecting the best fireworks for the next generation.	- N: number of locations - Number of iteration - Number of generating sparks	-A set of location of fireworks. -Binary, real, integer, structured tree	Business intelligence, data mining problem, network planning, scheduling problem, image segmentation, and electromagnetism and mechanics computational fluid.
CCA	- Initial displacement - Shelter darkness - Congeners attraction - The security quality - Permutation - Evaluation - Displacement and updating	- Number of initial cockroaches - Iteration number - Probability of changing shelter - Best quality security of each cockroaches in each shelter previously visited - The attraction rate of the neighbourhood	-Vector (binary, real, matrix) Structured tree	Spam detection, plagiarism detection, unsupervised classification, information retrieval, traveling salesman problem, knapsack problem, airéin routing problem, optemisation of multi-objective problem.
BA	- Initialization of echolocation parameters - initial displacement - Movement - local Search - Evaluation - Updating the parameters of echolocation	- The velocity (v_i). - The position (x_i). - A fixed minimum frequency (f_{min}). - A wavelength variable (). - intensity (Ao), At: is the average intensity of all bats.	-A set of artificial bats. -Vector (binary, real, matrix)	Machine vision system, image processing, decision support system, scheduling problem, the problem of wireless mobile networks.

biological mechanisms. The correct practice of bio-inspired artificial intelligence requires effort to extract the scientific principles of organic intelligence from data and theories provided by biologists and engineering efforts to translate these principles into algorithms and technologies.

This paper offers a trip on the theory and applications of meta-heuristic algorithms where we have compiled a comprehensive literature review on a range of natural computing algorithms. These are drawn from an evolutionary metaphor or natural and biological phenomena, including evolutionary algorithms (GAs, PG, SE, SD, CA), and swarm intelligence algorithms (PSO, ACO, BFA, WCA,CCA, BA, and FWA). The description of each algorithm is structured in 4 concepts (originally the inspiration, the summary, the general process). This work has allowed us to better understand the functioning of existing methods to help researchers to design a new bio-inspired method.

As just see throughout this paper, nature is a rich source of inspiration for solving real problems. Applications and the growth of these algorithms in recent years is very drastic summers when they applied to many optimization problems in computer networks, control systems, bioinformatics, game theory, biometrics systems power, image processing, industrial and engineering, parallel and distributed computing, robotics, economics and finance, forecasting problems, applications involving computer security.

REFERENCES

Anandaraman, C., Sankar, A. V. M., & Natarajan, R. (2012). A new evolutionary algorithm based

Bell, W. J., Roth, L. M., & Nalepa, C. A. (2007). *Cockroaches: ecology, behavior, and natural history*. JHU Press.

Benahmed, K., Merabti, M., & Haffaf, H. (2012). Inspired social spider behavior for secure wireless sensor networks. [IJMCMC]. *International Journal of Mobile Computing and Multimedia Communications*, *4*(4), 1–10.

Beni, G., & Wang, J. (1993). *Swarm intelligence in cellular robotic systems. InRobots and Biological Systems: Towards a New Bionics?* Springer Berlin Heidelberg.

Blackwell, T., & Bentley, P. J., Improvised music with swarms, In David B. Fogel, Mohamed A. El-Sharkawi, Xin Yao, Garry Greenwood, Hitoshi Iba, PaulMarrow, and Mark Shackleton (eds.), Proceedings of the 2002 Congress on Evolutionary Computation CEC 2002, pages 1462–1467, IEEE Press, 2002.

Bouarara, H. A., & Bouarara, Y. (2016). Swarm Intelligence Methods for Unsupervised Images Classification: Applications and Comparative Study. [IJOCI]. *International Journal of Organizational and Collective Intelligence*, *6*(2), 50–74.

Bouarara, H. A., Hamou, R. M., & Amine, A. (2014). Text Clustering using Distances Combination by Social Bees: Towards 3D Visualisation Aspect. [IJIRR]. *International Journal of Information Retrieval Research*, *4*(3), 34–53.

Bouarara, H. A., Hamou, R. M., & Amine, A. (2015). New Swarm Intelligence Technique of Artificial Social Cockroaches for Suspicious Person Detection Using N-Gram Pixel with Visual Result Mining. [IJSDS]. *International Journal of Strategic Decision Sciences*, *6*(3), 65–91.

Bouarara, H. A., Hamou, R. M., & Amine, A. (2015). Novel Bio-Inspired Technique of Artificial Social Cockroaches (ASC). [IJOCI]. *International Journal of Organizational and Collective Intelligence*, *5*(2), 47–79.

Bouarara, H. A., Hamou, R. M., & Amine, A. (2015). A Novel Bio-Inspired Approach for Multilingual Spam Filtering. [IJIIT]. *International Journal of Intelligent Information Technologies*, *11*(3), 45–87.

Bouarara, H. A., Hamou, R. M., Rahmani, A., & Amine, A. (2015). Boosting Algorithm and Meta-Heuristic Based on Genetic Algorithms for Textual Plagiarism Detection. [IJCINI]. *International Journal of Cognitive Informatics and Natural Intelligence*, *9*(4), 65–87.

Bourjot, C., Chevrier, V., & Thomas, V. (2003). A new swarm mechanism based on social spiders colonies: From web weaving to region detection. *Web Intelligence and Agent Systems: An International Journal*, *1*(1), 47–64.

Bremermann, H. (1974). Chemotaxis and optimization. *Journal of the Franklin Institute*, *297*(5), 397–404.

Colorni, A., Dorigo, M., Maniezzo, V., & Trubian, M. (1994). Ant System for Job-shop Scheduling. *Belgian Journal of Operations Research, Statistics and Computer Science*, *34*(1), 39–53.

Cramer, N. L. (1985, July). A representation for the adaptive generation of simple sequential programs. In *Proceedings of the First International Conference on Genetic Algorithms* (pp. 183-187).

Darwin, C. (1859). *On the origin ofspecies by means of natural selection, or the preservation of favoured races in the struggle for life*. D. Appleton and Company.

Darwin, C. (1991). *On the origin of species by means of natural selection. 1859*. Murray.

Darwin, C. (2009). The origin of species by means of natural selection: or, the preservation of favored races in the struggle for life.

Dorigo, M. Optimization, learning and natural algorithms (in Italian), Ph.D. Thesis, Dipartimento diElettronica, Politecnico di Milano, Italy, 1992.

Dorigo, M., & Gambardella, L. M. (1997). Ant colony system: A cooperative learning approach to the traveling salesman problem. Evolutionary Computation. *IEEE Transactions on*, *1*(1), 53–66.

Dorigo, M., & Gambardella, L. M. (1997). Ant colonies for the travelling salesman problem. *Bio Systems*, *43*(2), 73–81.

Dorigo, M., & Gambardella, L. M. (1997). Ant colony system: A cooperative learning approach to the traveling salesman problem. Evolutionary Computation. *IEEE Transactions on*, *1*(1), 53–66.

Dorigo, M., Maniezzo, V., & Colorni, A. Positive feedback as a search strategy, Tech. Report 91-016, Dipartimento di Elettronica, Politecnico di Milano, Italy, 1991.

Dorigo, M., Maniezzo, V., & Colorni, A. (1996). Ant System: Optimization by a colony of cooperating agents. *IEEE Transactions on Systems, Man, and Cybernetics. Part B, Cybernetics*, *26*, 29–41.

Dorigo, M., Maniezzo, V., & Colorni, A. (1996). Ant system: optimization by a colony of cooperating agents. Systems, Man, and Cybernetics, Part B: Cybernetics. *IEEE Transactions on*, *26*(1), 29–41.

Eberhart, R. C., & Shi, Y. Tracking and optimizing dynamic systems with particle swarms, *Proc. Congress on Evolutionary Computation 2001*, Seoul, Korea, 2001

Engelbrecht, A. P. (2007). *Computational intelligence: an introduction*. John Wiley & Sons.

Feng, X., Lau, F. C., & Gao, D. (2009). A new bio-inspired approach to the traveling salesman problem. In *Complex Sciences* (pp. 1310–1321). Springer Berlin Heidelberg.

Fuller, T. K., Mech, L. D., & Cockrane, J. F. (2003). Wolf population dynamics. In L. D. Mech & L. Boitani (Eds.), *Wolves: Behavior, Ecology and Conservation* (pp. 161–191). University of Chicago Press.

Gandomi, A. H., Yang, X. S., & Alavi, A. H. (2013). Cuckoo search algorithm: A metaheuristic approach to solve structural optimization problems. *Engineering with Computers*, *29*(1), 17–35.

Grassé, P. P. (1967). Nouvelles experiences sur le termite de Müller (Macrotermes mülleri) et considerations sur la théorie de la stigmergie. *Insectes Sociaux*, *14*(1), 73–101.

Hicklin, J. F. (1986). Application of the genetic algorithm to automatic program generation.

Hofstadter, D. (2013). *Alan Turing: Life and legacy of a great thinker* (C. Teuscher, Ed.). Springer Science & Business Media.

Holland, J. H. (1975). *Adaptation in natural and artificial systems*. The University of Michigan Press.

Janecek, A., & Tan, Y. (2011, July). Iterative improvement of the multiplicative update nmf algorithm using nature-inspired optimization. In Natural Computation (ICNC), 2011 Seventh International Conference on (Vol. 3, pp. 1668-1672). IEEE. doi:10.1109/ICNC.2011.6022356

Kennedy, J., & Eberhart, R. C. Particle Swarm Optimization. In *Proceedings of IEEE International Conference on Neural Networks*, Perth, Australia, pp. 1942–1948, 1995

Koza, J. R. (1992). *Genetic programming: on the programming of computers by means of natural selection* (Vol. 1). MIT press.

Koza, J. R. (1992). *Genetic programming: on the programming of computers by means of natural selection* (Vol. 1). MIT press.

Koza, J. R. (1994). Genetic programming II: Automatic discovery of reusable subprograms. Cambridge, MA, USA.

Li, J., Zheng, S., & Tan, Y. (2014, July). Adaptive Fireworks Algorithm. In Evolutionary Computation (CEC), 2014 IEEE Congress on (pp. 3214-3221). IEEE. doi:10.1109/CEC.2014.6900418

Li, W.-W., Wang, H., & Zou, Z. J. (2005). Function optimization method based on bacterial colony chemotaxis. *Journal of Circuits and Systems*, *10*, 58–63.

Liu, C., Yan, X., Liu, C., & Wu, H. (2011). The wolf colony algorithm and its application. *Chinese Journal of Electronics*, *20*, 212–216.

Liu, J., Zheng, S., & Tan, Y. (2013). The improvement on controlling exploration and exploitation of firework algorithm. In *Advances in swarm intelligence* (pp. 11–23). Springer Berlin Heidelberg., doi:10.1007/978- 3-642-38703-6_2

Macdonald, D. W., Creel, S., & Mills, M. G. L. (2004). Society: Canid society. In D. W. Macdonald & C. Sillero-Zubiri (Eds.), *Biology and Conservation of Wild Carnivores* (pp. 85–106). Oxford University Press.

Mitchell, M. (1998). *An introduction to genetic algorithms*. MIT press.

Niu, B., & Wang, H. (2012). *Bacterial colony optimization*. Discrete Dynamics in Nature and.

on bacterial evolution and its applications for scheduling a flexible manufacturing system. Jurnal Teknik Industri, 14, 1–12

Passino, K. M. (2012). Bacterial foraging optimization. Innovations and Developments of Swarm Intelligence Applications, 219.

Pei, Y., Zheng, S., Tan, Y., & Takagi, H. (2012, October). An empirical study on influence of approximation approaches on enhancing fireworks algorithm. In Systems, Man, and Cybernetics (SMC), 2012 IEEE International Conference on (pp. 1322-1327). IEEE. doi:10.1109/ICSMC.2012.6377916

Pohlheim, H. (2005). Evolutionary algorithms: overview, methods and operators. GEATbx: Gentic & Evolutionary AlgorithmToolbox for Matlab.

Poli, R., Langdon, W. B., McPhee, N. F., & Koza, J. R. (2008). A field guide to genetic programming. Lulu. com.

Premaratne, U., Samarabandu, J., & Sidhu, T. (2009, December). A new biologically inspired optimization algorithm. In Industrial and Information Systems (ICIIS), 2009 International Conference on (pp. 279-284). IEEE.

Rechenberg, I. (1989). *Evolution strategy: Nature's way of optimization. InOptimization: Methods and applications, possibilities and limitations*. Springer Berlin Heidelberg.

Reeves, C. (2003). *Genetic algorithms*.

Reynolds, C. W. (1987, August). Flocks, herds and schools: A distributed behavioral model. [). ACM.]. *Computer Graphics*, *21*(4), 25–34.

Runkler, T. A. (2008). Wasp swarm optimization of the c-means clustering model. *International Journal of Intelligent Systems*, *23*(3), 269–285.

Storn, R., & Price, K. (1997). Differential evolution–a simple and efficient heuristic for global optimization over continuous spaces. *Journal of Global Optimization*, *11*(4), 341–359.

Stützle, T., & Hoos, H. (1997, April). MAX-MIN ant system and local search for the traveling salesman problem. In Evolutionary Computation, 1997., IEEE International Conference on (pp. 309-314). IEEE.

Wallace, A. R. (1870). *Contributions to the theory of natural selection: A series of essays*. Macmillan and Company.

Yang, C., Tu, X., & Chen, J. (2007). Algorithm of marriage in honey bees optimization based on the wolf pack search. In *IEEE International Conference on Intelligent Pervasive Computing (IPC)* (pp. 462–467)

Yoshida, H., Kawata, K., Fukuyama, Y., Takayama, S., & Nakanishi, Y. (2000). A particle swarm optimization for reactive power and voltage control considering voltage security assessment. *IEEE Transactions on Power Systems*, *15*(4), 1232–1239.

Zheng, S., Janecek, A., Li, J., & Tan, Y. (2014, July). Dynamic search in fireworks algorithm. In Evolutionary Computation (CEC), 2014 IEEE Congress on (pp. 3222-3229). IEEE. doi:10.1109/CEC.2014.6900485

Zheng, S., Janecek, A., & Tan, Y. (2013, June). Enhanced fireworks algorithm. In Evolutionary Computation (CEC), 2013 IEEE Congress on (pp. 2069-2077). IEEE. doi:10.1109/CEC.2013.6557813

Chapter 3
Health Precautions for Patient Safety

Cheimaa Bentadj
Dr. Moulay Tahar University, Algeria

ABSTRACT

In December 2019, an outbreak of severe acute respiratory syndrome coronavirus 2 (SARSCoV-2) infection occurred in Wuhan, Hubei Province, China and spread across China and beyond. On February 12, 2020, WHO officially named the disease caused by the novel coronavirus as Coronavirus Disease 2019. In the literature, COVID-19 is characterized by the symptoms of viral pneumonia such as fever, fatigue, dry cough, and lymphopenia. Many of the older patients who become severely ill have evidence of underlying illness such as cardiovascular disease, liver disease, kidney disease, or malignant tumors. The COVID-19 pandemic posed a great danger at all levels, and despite the doctors' rush to find a vaccine, it will remain a threat to the world unless people adhere to health precaution. The discussion in this chapter of health precautions will focus largely on two main components: wearing face masks and respecting social distancing, which is the distance regulated by the World Health Organization (1 meter); also it will discusses the symptoms caused by this pandemic and its impact on some areas.

INTRODUCTION

In December 2019, an outbreak of severe acute respiratory syndrome coronavirus 2 (SARSCoV-2) infection occurred in Wuhan, Hubei Province, China and spread across China and beyond. On February 12, 2020, WHO officially named the disease caused by the novel coronavirus as Coronavirus Disease 2019 (COVID-19) (Zu,

DOI: 10.4018/978-1-6684-5656-9.ch003

2020). Since most COVID-19 infected patients were diagnosed with pneumonia and characteristic CT imaging patterns, radiological examinations have become vital in early diagnosis and assessment of disease course. To date, CT findings have been recommended as major evidence for clinical diagnosis of COVID-19 in Hubei, China. This review focuses on the etiology, epidemiology, and clinical symptoms of COVID-19, while highlighting the role of chest CT in prevention and disease (Zu, 2020).

COVID 19 PANDEMIC

COVID-19 is the disease caused by a new coronavirus called SARS-CoV-2. WHO first learned of this new virus on 31 December 2019, following a report of a cluster of cases of 'viral pneumonia' in Wuhan, People's Republic of China. In most cases, COVID-19 causes mild symptoms including dry cough, tiredness and fever, though fever may not be a symptom for some older people. Other mild symptoms include aches and pains, nasal congestion, runny nose, sore throat or diarrhoea. Some people become infected but don't develop any symptoms and don't feel unwell. Most people recover from the disease without needing special treatment. Around 1 out of every 6 people who get COVID-19 becomes seriously ill and has difficulty breathing.

MAIN SYMPTOMS OF CORONAVIRUS COVID 19

High Temperature

World Health Organization (WHO) reported that there have been pneumonia cases in Wuhan City, Hubei Province, China, but the etiology was unknown. The case was developed very fast, until January 7, 2020, the Chinese government said that pneumonia was a new type of coronavirus or covid-19. Common signs and symptoms of covid-19 infection include symptoms of acute respiratory disorders such as fever, coughing and shortness of breath.

The average incubation period is 5–6 days with the longest incubation period of 14 days. In severe cases, covid-19 can cause pneumonia, acute respiratory syndrome, kidney failure, and even death. The clinical signs and symptoms reported in the majority of cases are fever, with some cases having difficulty breathing, and X-rays show extensive pneumonia infiltrates in both lungs (Tosepu, 2020).

Figure 1. Situation by WHO Region to the number of confirmed cases of COVID-19 in the world

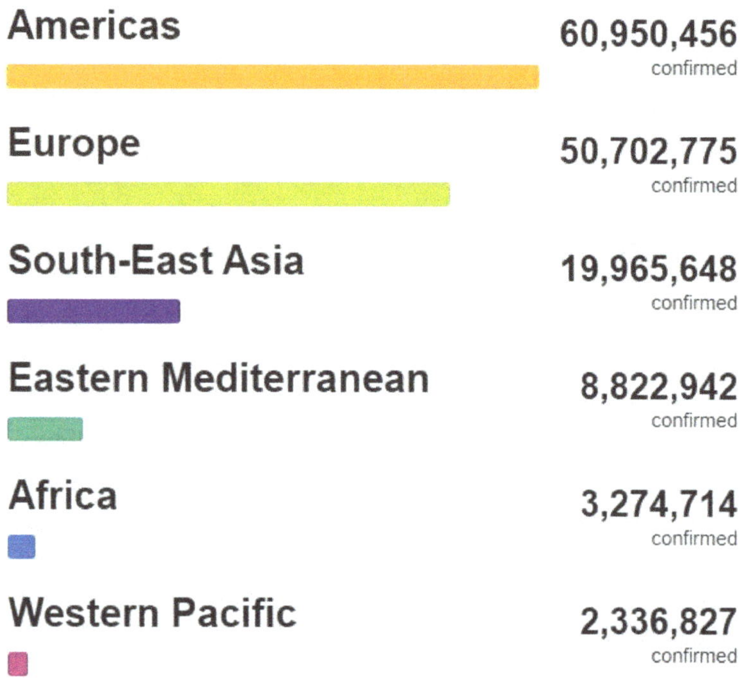

Effects of COVID-19 Pandemic

COVID-19 (Coronavirus) has affected day to day life and is slowing down the global economy. This pandemic has affected thousands of peoples, who are either sick or are being killed due to the spread of this disease. The most common symptoms of this viral infection are fever, cold, cough, bone pain and breathing problems, and ultimately leading to pneumonia. This, being a new viral disease affecting humans for the first time, vaccines are not yet available. Thus, the emphasis is on taking extensive precautions (Haleem, 2020), such as social distancing, and wearing of masks and this is among what we will cover in this thesis.

Presently, the impacts of COVID-19 are widespread and have far-reaching consequences. It can be divided into different categories:

- Healthcare
    - Challenges in the diagnosis, quarantine and treatment of suspected or confirmed cases
    - High burden of the functioning of the existing medical system

- Patients with other disease and health problems are getting neglected
- Overload on doctors and other healthcare professionals, who are at a very high risk
- Overloading of medical shops
- Requirement for high protection
- Disruption of medical supply chain

- Economic
  - Slowing of the manufacturing of essential goods
  - Disrupt the supply chain of products
  - Losses in national and international business
  - Poor cash flow in the market
  - Significant slowing down in the revenue growth
- Social
  - Service sector is not being able to provide their proper service
  - Cancellation or postponement of large-scale sports and tournaments
  - Avoiding the national and international travelling and cancellation of services
  - Disruption of celebration of cultural, religious and festive events
  - Undue stress among the population
  - Social distancing with our peers and family members
  - Closure of the hotels, restaurants and religious places
  - Closure of places for entertainment such as movie and play theatres, sports clubs, gymnasiums, swimming pools, and so on.
  - Postponement of examinations

This COVID-19 has affected the sources of supply and effects the global economy. There are restrictions of travelling from one country to another country. During travelling, numbers of cases are identified positive when tested, especially when they are taking international visits. All governments, health organizations and other authorities are continuously focussing on identifying the cases affected by the COVID-19. Healthcare professional face lot of difficulties in maintaining the quality of healthcare in these days (Haleem, 2020).

HOW TO PROTECT YOURSELF AND OTHERS

Wearing a Mask

The World Health Organization (WHO) advises the use of masks as part of a comprehensive package of prevention and control measures to limit the spread of

SARS-CoV-2, the virus that causes COVID-19 (World Health Organization, 2020), everyone 2 years and older should wear masks in public, masks should be worn in addition to staying at least 6 feet apart, especially around people who don't live with you, if someone in your household is infected, people in the household should take precautions including wearing masks to avoid spread to others. WHO continues to advise that anyone suspected or confirmed of having COVID-19 or awaiting viral laboratory test results should wear a medical mask when in the presence of others (this does not apply to those awaiting a test prior to travel). Depending on the type, masks can be used either for protection of healthy persons or to prevent onward transmission (source control), for any mask type, appropriate use, storage and cleaning or disposal are essential to ensure that they are as effective as possible and to avoid an increased transmission risk.

Social Distancing

However, the use of a mask alone is insufficient to provide an adequate level of protection or source control, and other personal and community level measures should also be adopted to suppress transmission of respiratory viruses. Whether or not masks are used, compliance with social distancing and other infection prevention and control (IPC) measures are critical to prevent human-to human transmission of COVID-19 (World Health Organization, 2020).

Social distancing refers to measures aimed at reducing interactions within a community, which can include infected individuals as yet unidentified, hence not in isolation. Since diseases transmitted through respiratory droplets require a certain physical proximity for contagion to occur, social distancing allows transmission to be reduced. Examples of social distancing measures that have been adopted include: the closure of schools and workplaces, closure of certain businesses, and cancellation of events to avoid mass gatherings. Social distancing is particularly useful in settings where there is community transmission of the virus, where the restriction measures imposed exclusively on known cases or on the most vulnerable segments of the population are considered insufficient to prevent new transmissions (Wilder-Smith A, 2019).

COVID-19 VACCINE

The world is in the midst of a COVID-19 pandemic. As WHO and partners work together on the response tracking the pandemic, advising on critical interventions, distributing vital medical supplies to those in need they are racing to develop and deploy safe and effective vaccines. Vaccines save millions of lives each year. Vaccines work by training and preparing the body's natural defenses the immune system to recognize and fight off the viruses and bacteria they target. After vaccination, if the body is later exposed to those disease-causing germs, the body is immediately ready to destroy them, preventing illness. Vaccines are a critical new tool in the battle against COVID-19 and it is hugely encouraging to see so many vaccines proving successful and going into development. Working as quickly as they can, scientists from across the world are collaborating and innovating to bring us tests, treatments and vaccines that will collectively save lives and end this pandemic.

There are three main approaches to designing a vaccine:

- The whole-microbe approach
- The subunit approach
- The genetic approach (nucleic acid vaccine)

As of 18 February 2021, at least seven different vaccines across three platforms have been rolled out in countries. Vulnerable populations in all countries are the highest priority for vaccination. At the same time, more than 200 additional vaccine candidates are in development, of which more than 60 are in clinical development. COVAX is part of the ACT Accelerator, which WHO launched with partners in 2020. COVAX, the vaccines pillar of ACT Accelerator, convened by CEPI, Gavi and WHO, aims to end the acute phase of the COVID-19 pandemic by[1]:

- Speeding up the development of safe and effective vaccines against COVID-19,
- Supporting the building of manufacturing capabilities,
- Working with governments and manufacturers to ensure fair and equitable allocation of the vaccines for all countries – the only global initiative to do so[1].

Vaccine Doses

For some COVID-19 vaccines, two doses are required. It's important to get the second dose if the vaccine requires two doses. For vaccines that require two doses, the first dose presents antigens – proteins that stimulate the production of antibodies – to the immune system for the first time. Scientists call this priming the immune response. The second dose acts as a booster, ensuring the immune system develops a memory response to fight off the virus if it encounters it again.

Because of the urgent need for a COVID-19 vaccine, initial clinical trials of vaccine candidates were performed with the shortest possible duration between doses. Therefore an interval of 21–28 days (3–4 weeks) between doses is recommended

Figure 2. tables of COVID-19 vaccine candidates in both clinical and pre-clinical development[1]

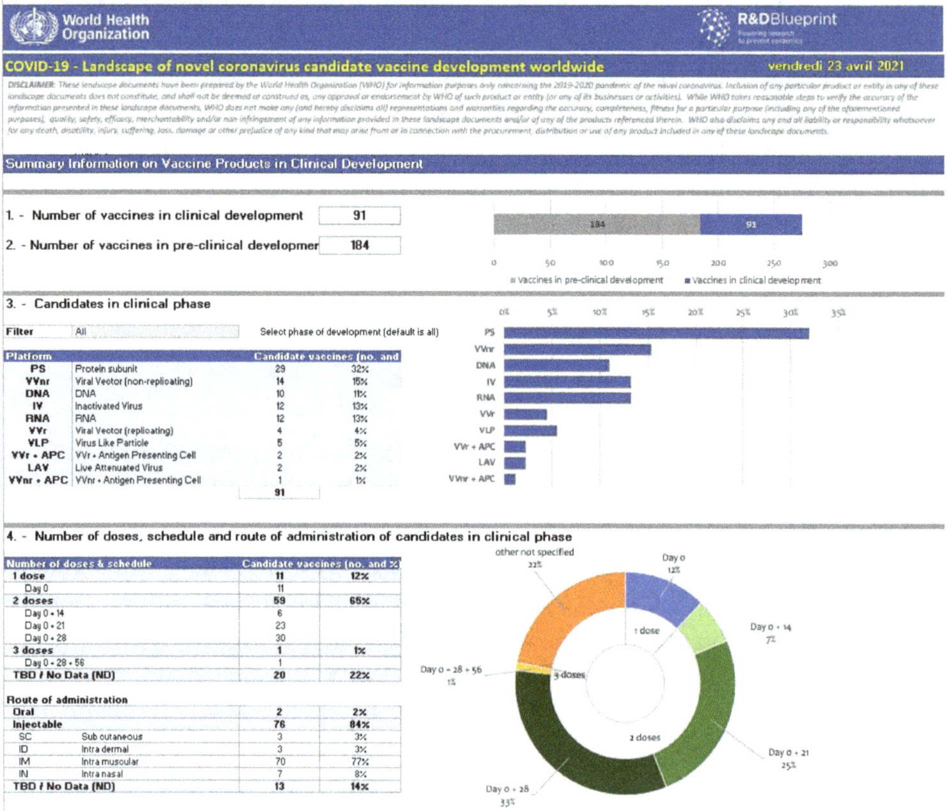

by WHO. Depending on the vaccine, the interval may be extended for up to 42 days – or even up to 12 weeks for some vaccines – on the basis of current evidence[1].

The COVID-19 Candidate Vaccine Landscape and Tracker

provides summary tables of COVID-19 vaccine candidates in both clinical and pre-clinical development, provides analysis and visualization for several COVID-19 vaccine candidate categories, tracks the progress of each vaccine from pre-clinical, Phase 1, Phase 2 through to Phase 3 efficacy studies and including Phase 4 registered as interventional studies, provides links to published reports on safety, immunogenicity and efficacy data of the vaccine candidates, includes information on key attributes of each vaccine candidate and allows users to search for COVID-19 vaccines through various criteria such as vaccine platform, schedule of vaccination, route of administration, developer, trial phase and clinical endpoints[1].

CONCLUSION

Coronavirus is an infectious disease that is transmitted from animals to humans and has caused many problems at all social, economic and health levels, and to reduce this crisis, the World Health Organization has suggested some measures to prevent it and scientists also from across the world are collaborating and innovating to bring us tests, treatments and vaccines that will collectively save lives and end this pandemic.

Vaccines are a critical new tool in the battle against COVID-19 and it is hugely encouraging to see so many vaccines proving successful and going into development.

REFERENCES

Haleem, A., Javaid, M., & Vaishya, R. (2020). Effects of COVID-19 pandemic in daily life. *Current Medicine Research and Practice, 10*(2), 78.

Tosepu, R., Gunawan, J., Effendy, D. S., Lestari, H., Bahar, H., & Asfian, P. (2020). Correlation between weather and Covid-19 pandemic in Jakarta, Indonesia. *The Science of the Total Environment, 725*, 138436. doi:10.1016/j.scitotenv.2020.138436 PMID:32298883

Wilder-Smith, A., & Freedman, D. O. (2020). Isolation, quarantine, social distancing and community containment: Pivotal role for old-style public health measures in the novel coronavirus (2019-nCoV) outbreak. *Journal of Travel Medicine, 27*(2), 2. doi:10.1093/jtm/taaa020 PMID:32052841

World Health Organization. (2020). Advice on the use of masks in the context of COVID-19: interim guidance, 5 June 2020 (No. WHO/2019-nCoV/IPC_Masks/2020.4). World Health Organization.

Zu, Z. Y., Jiang, M. D., Xu, P. P., Chen, W., Ni, Q. Q., Lu, G. M., & Zhang, L. J. (2020). Coronavirus disease 2019 (COVID-19): A perspective from China. *Radiology, 296*(2), E15–E25. doi:10.1148/radiol.2020200490 PMID:32083985

ENDNOTE

1 World Health Organization (2021).

Chapter 4
Homology Modeling and Binding Site Analysis of SARS-CoV-2 (COVID-19) Main Protease 3D Structure

Marion Olubunmi Adebiyi
https://orcid.org/0000-0001-7713-956X
Landmark University, Omuo-Aran, Nigeria

Ibidun Christiana Obagbuwa
https://orcid.org/0000-0002-7965-2823
Sol Plaajte University, Kimberley, South Africa

ABSTRACT

The severe acute respiratory syndrome coronavirus 2 (SAR-Cov-2) caused the coronavirus (COVID-19) pandemic. The global concern is the discovery of a new target drug for the total cure. Recently, some research showed that a few COVID-19 enzymes may have been contemplated to be potential drug targets, but not much is known about its structural biology. This research investigates the 3-D structure of protease SAR-CoV-2. The tertiary structure was determined by homology modeling. The Swiss-Model workspace and the basic local alignment search tool (BLAST) were employed for modeling, and the resulted model was validated with programs that include PROCHECK, Verify3D, and QMEAN to ascertain reliability. To establish the structures that fitted, HHBlits software was employed. To verify the structure quality, a Ramachandran plot was exploited. The binding site was determined using CASTp and DeepSite algorithms, which showed that this protein can be exploited as a prospective pharmaceutical target. Albeit further experimentation is required as a COVID-19 virus vaccine-design/target-drug.

DOI: 10.4018/978-1-6684-5656-9.ch004

1. INTRODUCTION

The recent global outbreak disease called coronavirus 19 (COVID-19) with high sickness and death rates is induced by the severe acute respiratory syndrome coronavirus 2 (SAR-Cov-2). New distress from the Huanan seafood market, Wuhan, China, surfaced in December 2019 (Fox, 2020; Chellapandi, P., & Saranya, S.,2020). By 11 March, the world health organization (WHO) enunciated that the eruption was a pandemic, and by the 20th of July 2020, 14.6 million cases were affirmed in over 210 countries across the globe with 608 deaths. Nations such as the USA, Italy, and Germany have recorded more cases than China. To quickly coordinate the scientific and medical procedures to develop a cure, the WHO announces a state of a global health emergency (see- https://www.worldometers.info/coronavirus/). The world is craving vaccines, but little operative target drug has been feasible recently, and further investigation is ongoing. The pathogenesis techniques, transmission, and source of COVID-19 are still unknown, which impedes the investigation of working antiviral (Saranya, S.,2020). Nevertheless, one of the breakthroughs for this challenge is identifying the protease site of the SAR-Cov-2 and obstructing the metamorphoses that cause rapid transmission of the protection of the infected individuals. The immense investigations on COVID-19 genomics offer assurance of finding the vaccines to prevent and cure the virus.

Homology modeling is considered a vital technique in structural biology for reducing disparities between known protein sequences and experimentally determined structures. This includes the amino acid sequences of the interacting proteins, stoichiometry, and the whole structure of the complex. Many insights taken from the protein's 3D structure provide helpful information about their function from a molecular point of view and a wide range of applications in life science research. Although protein complexes are central to several cellular procedures, the entire quaternary structure is essential for the clear expression of the biological system.

In this work, the 3D structure of a SARS-CoV-2 (COVID-19) Main Protease bound to 2-Methyl-1-tetralone (PEPTIDE BINDING PROTEIN) (Guenther et al., 2020) is reported. Besides, the computer simulation of the prediction of its active site for a potential drug target is also provided.

2. MATERIALS AND METHODS

2.1 Computational Methods

The 3D modeling computational methods embroiled template selection, template alignment with the target, building model, and structure evolution. SARS-CoV-2

Main Protease sequence was fetched from the protein database (PDB) (Guenther et al.)- https://rcsb.org/structure/6ynq.

2.2. Searching for Template

BLAST and HHBlits were used for template searching in conjunction with the SWISS-MODEL template library (SMTL, last update: 2020-11-11, last included PDB release: 2020-11-06).

BLAST was engaged for the search for target sequence with the content of Amino acids in the SMTL and found 165 templates. The procedures listed in (Steinegger et al.) were used to build the initial HHblists profile, followed by one iteration of HHblits against Uniclust30 (von den Driesch et al.). The obtained profile has then been searched against all profiles of the SMTL. A total of 480 templates were found.

2.3 Template Selection and Sequence Alignment

The basic local alignment tool (BLAST) server was used to thoroughly examined the protein data bank (PDB) and the UniProt Knowledgebase (UniProtKB)-accession number: P0DTD1 to discover the similar homologs of the sequence being enquired. 6YNQ_1|Chain A|Replicase polyprotein 1ab|severe acute respiratory syndrome coronavirus 2 (2697049) with RID (VP2K8RUB01R) carried out, and the discovered best template has accession number QOQ24800.1 (Stephen et al., 1997). The template quality was predicted from target template alignment features, and the highest quality template was selected for building the model.

2.4 Model Building

The SARS-CoV-2 Main Protease model PDB file was built using CPH models-3.2 Server (Nielsen et al., 2010), Swiss model workspace was utilized to derive the 3D structure (Waterhouse et al., Guex et al., Bienert et al.). The SWISS-MODEL template library (SMTL version 2020-11-11, PDB release 2020-11-06) was searched with BLAST (Camacho et al.) and HHBlits (Steinegger et al.) for related evolutionary structures matching the target sequence. ProMod3 was used to build models on the target template. Coordinates between the target and the template are copied from the template to the model. A fragment library was used to remodel insertion and deletion, and sidechains were built again. Finally, the geometry of the resulting model is regularized by using a force field. In case of loop modeling with ProMod3 fails, an alternative model is built with PROMOD-II (Guex et al.).

2.5 Evolution and Validation of a Model

The best model was achieved using the program PROCHECK (Laskowski et al.) to validate the stereochemical quality of the models. Besides, the Verify3D structure evaluation server was used for the 3D-profiling of the residue (Bowie et al., 1991, Luethy et al., 1992), and the QMEAN scoring function was utilized for assessing the global and per-residue model quality (Studer et al.).

2.6 Secondary Structure and Active Site Analysis

The self-optimized prediction method with alignment (SOPMA) was used for the secondary structure evolution (Combet et al, 2000). Moreover, the active site analysis was accomplished with DeepSite and Computed Atlas of Surface Topography of proteins (CASTp) for the modeled protein and its docking studies. It made a grid before docking is given by the active site analysis (Tian et al., 2018).

2.7 Ligand Modelling

Ligands present within the template structure are transferred by the homology to the model when the subsequent criteria are met: (a) The ligands are annotated as biologically relevant within the template library, (b) the ligand is in touch with the model, (c) the ligand isn't clashing with the protein, (d) the residues in touch with the ligand are conserved between the target and therefore the template. If any of those four criteria aren't satisfied, a particular ligand will not be included in the model. The information on why and which ligand has not been included is provided by the model summary.

2.8 Oligomeric State Conservation

The quaternary structure annotation of the template is employed to model the target sequence in its oligomeric form. The tactic (Bertoni et al.) is predicated on a supervised machine learning algorithm, Support Vector Machines (SVM), which mixes interface conservation, structural clustering, and other template features to supply a quaternary structure quality estimate (QSQE). The QSQE score may be a number between 0 and 1, reflecting the expected accuracy of the interchain contacts for a model built supported a given alignment and template. Higher numbers indicate higher reliability. This complements the GMQE score, which estimates the accuracy of the tertiary structure of the resulting model.

3. RESULTS

3.1 Template Selection and Sequence Alignment

The structure of the COVID-19 main protease has been built using the template sequence of chain A, SARS-CoV-2 (COVID-19). Main Protease is presented in Figure 1.

3.2 Model-3D Structure

The 3D structure of COVID-19 main protease presented was built by homology modeling using the template collected from the protein data bank. The 3D model is depicted in Table1 (Model #01). Furthermore, Tables 1, 2, and 3 illustrate the model built from the target sequence, selected template, and search information.

Table 1. Model built from the target sequence

Model #01	File	Built with	Oligo-State	Ligands	GMQE	QMEAN
	PDB	ProMod3 3.1.1	homo-dimer (matching prediction)	2 x JRY: hexylcarbamic acid;	0.97	0.72

Table 2. Selected Template

Template	Seq Identity	Oligo-state	QSQE	Found by	Method	Resolution	Seq Similarity	Range	Coverage	Description
7buy.1.A	100.00	homo-dimer	1.00	HHblits	X-ray	1.60Å	0.63	1 - 306	1.00	SARS-CoV-2 virus Main protease

Table 3. Details on the template search

Template	Seq Identity	Oligo-state	QSQE	Found by	Method	Resolution	Seq Similarity	Coverage	Description
7buy.1.A	100.00	homo-dimer	1.00	HHblits	X-ray	1.60Å	0.63	1.00	SARS-CoV-2 virus Main protease
6wtk.1.B	100.00	homo-dimer	0.99	HHblits	X-ray	2.00Å	0.63	1.00	3C-like proteinase
6wtj.1.B	100.00	homo-dimer	0.99	HHblits	X-ray	1.90Å	0.63	1.00	3C-like proteinase
7c8r.1.B	100.00	homo-dimer	1.00	HHblits	X-ray	2.30Å	0.63	1.00	3C-like proteinase
6y2g.1.B	100.00	homo-dimer	1.00	HHblits	X-ray	2.20Å	0.63	1.00	Replicase polyprotein 1ab
6yz6.1.A	100.00	homo-dimer	0.98	HHblits	X-ray	1.70Å	0.63	1.00	Main Protease
5rgo.1.A	100.00	homo-dimer	0.96	HHblits	X-ray	1.74Å	0.63	1.00	3C-like proteinase
6yvf.1.B	100.00	homo-dimer	0.95	HHblits	X-ray	1.60Å	0.63	1.00	Replicase polyprotein 1ab

3.3 Model Validation

3.3.1 Validation of the Structure

The Ramachandran plot of non-glycine and non-proline residues distribution values are illustrated in Figure 2 (a) & (b).

3.3.2 3D Model Verification

The verify3D graph depicted in Figure 3 shows a profile score above zero, which matches the suitable environment of the model. Figure 4(b) presents the target sequence global quality model measured by the QMEAN6 score based on the linear combination of six structural descriptors. Figure 4(b) reflects the reliability of the predicted model ranging from 0 to 1 (Benkert et al.). The raw QMEAN6 score and average Z-score for COVID-19 main protease are depicted in Figure 4(a).

3.4 Secondary Structure and Active Site of COVID-19 Main Protease

3.4.1 Analysis of the Secondary Structure

The secondary structure analysis is used to determine the position of a given amino acid, which can be in a helix strand or coil. Table 4 and Figure 5 show the secondary structural features predicted using SOPMA with default parameters- window width: 17, similarity threshold: 8, and the number of states: 4.

3.4.2 Active Site Determination

COVID-19 Main protease active site was predicted using CASTp server 3.0, as presented in Figure 6. Moreover, the best active site area of the experimental enzyme and the number of amino acids contained are shown in Figure 6b. Further analysis of the active site was done using the DeepSite server (Figure 7).

4. DISCUSSION

The homology modeling algorithm is one of the most popular and highly robust bioinformatics modeling tools. Many research has shown it to be a good tool for effective and efficient modeling. The chain A template sequence of the primary amino acid sequence shown in Figure 1 was used to build the COVID-19 main protease structure. Table 1 depicts the summary of the model built from the target sequence, which includes the 3D structure of COVID-19 main protease by comparative modeling, the SARS-CoV-2 Main Protease PDB file. Models are built with ProMod3-3.1.1 based on the target-template alignment with 100% sequence identity, 1.0 coverage with homodimer (matching prediction) oligo-state 1.60A resolution and 0.63 sequence identity. Table 2 shows the selected template (PDB ID: 7buy.1. A) and its specifics. The best template was chosen from a long list of

Figure 1. COVID-19 Primary amino acid sequence (Target Sequence)

```
>6YNQ_1|Chain A |Replicase polyprotein 1ab|severe acute respiratory syndrome coronavirus 2 (2697049)
SGFRKMAFPSGKVEGCMVQVTCGTTTLNGLWLDDVVYCPRHVICTSEDMLNPNYEDLLIRKSNHNFL
VQAGNVQLRVIGHSMQNCVLKLKVDTANPKTPKYKFVRIQPGQTFSVLACYNGSPSGVYQCAMRPNF
TIKGSFLNGSCGSVGFNIDYDCVSFCYMHHMELPTGVHAGTDLEGNFYGPFVDRQTAQAAGTDTTITV
NVLAWLYAAVINGDRWFLNRFTTTLNDFNLVAMKYNYEPLTQDHVDILGPLSAQTGIAVLDMCASLK
ELLQNGMNGRTILGSALLEDEFTPFDVVRQCSGVTFQ
```

templates searched, as illustrated in Table 3. The top 8 filtered templates are shown in Table 3. However, other 471 templates were also discovered but considered not too suitable for modeling than the sifted list (Table 3).

The structure validation server verify3D was employed to verify the target sequence 3D structure model. The best structure was chosen based on the scores. The model is shown in Figure 3 to have an excellent environmental profile with a high score of 0.77 (Premalatha et al.). The allocation of a structural class and the position and surroundings- alpha, beta, loop, polar, nonpolar with the results compared to suitable structures is evidence that the 3D atomic model is compatible with its amino acids sequence (1D). The compatibility level of the COVID-19 model illustrated in the verify3D graph (Figure 3) affirms that the model has a PASS grade proportional to the model suitable environment with a rating of 92.16% of the residues have averaged 3D-1D score >= 0.2, indicating that at least 80% of the amino acids have scored >= 0.2 in the 3D/1D profile. The usage of the standard procedure in this work for constructing protein model show that the SARS-CoV-2 main protease contains alpha-helix, beta-sheets (Table 4 and Figure 6(e)), and 99.6% of amino acids in the favorable and allowed regions and 0.4% in the disallowed region presented in Figure 2 (a) & (b). The regions classified the residues in quadrangle according to the Ramachandran plot analysis. The most allowed regions are depicted in the red area, while the allowed regions are shown in the yellow zone. Glycine is denoted by triangles, whereas other residues are represented by squares, as illustrated in Figure 2(a). The PROCHECK analysis done in this work affirms that all the residues are within the boundary of the Ramachandran plot. This is evidence that the model is typical.

The QMEAN scoring function approximates the global quality of the models based on a linear combination of six structural descriptors. The four of them are the statistical potentials of the mean force. Figure 4 (a) & (b) illustrate the results of the QMEAN6, which was employed for the computation of the global model quality estimate (GMQE) of the target sequence. The local geometry is analyzed by a torsion angle potential over three consecutive amino acids. Two distance-dependent interaction potentials based on Cβ atoms and all atoms are used to assess long-range interactions (Benkert et al.). A solvation potential portrays the burial status of the residues. Cheng et al. reported two terms reflecting the agreement between the predicted and calculated secondary structure and solvent accessibility. The QMEAN Z-score measures the features of the degree of naturalness shown in a structural model, which illustrates that the model is equivalent to the existing experimental structures (Benkert et al.). According to the analysis in Figures 4 (a) & (b), the average Z-score for the COVID-19 main protease 3D-model is 0.72.

Figure 2. (a) Ramachandran Plot of SARS-CoV-2 Main Protease (b) Ramachandran Plot statistics of SARS-CoV-2 Main Protease

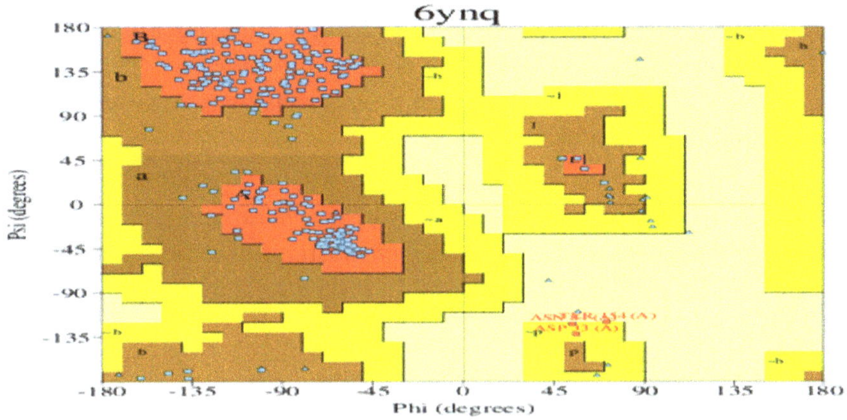

a

Statistics	Residue Number	Percentage (%)
Residues in the most favored regions (A, B, L)	235	88.7
Residues in the additional allowed regions (a, b, l, p)	27	10.2
Residues in the generously allowed regions (-a, -b, -l, -p)	2	0.8
Residues in the disallowed regions	1	0.4
Number of Non-glycine and non-proline residues	265	100
Number of End-residues (excl. Gly and Pro)	2	-
Number of Glycine residues	26	-
Number of Proline residues	13	-
Total number of residues	306	-

b

Figure 3. Verify3D Graph of SARS-CoV-2 Main Protease Model

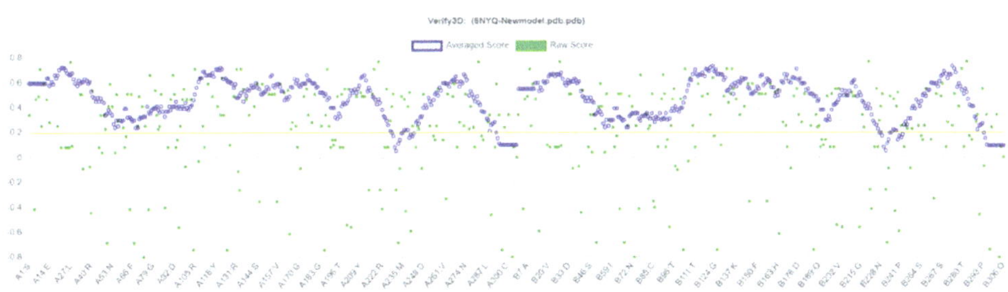

The GMQE report shows a raw score of 0.97 for the predicted COVID-19 main protease (Figure 4(a). The reliability scores of the GMQE metrics range from zero to one for the predicted model (Benkert et al.). The model is confirmed to be of good quality as depicted in Figure 4(a) & (b) with Z-scores of 0.49, -0.39, -0.78, and

Figure 4. (a) Global Model Quality Estimate Graph with GMQE (0.97) (b) Global Quality Estimate (Z-score) of the individual component of COVID-19 main protease. (c) Model Template Alignment (d): Local Quality Estimate Graph with Predicted Local Similarity Target (1.0) (e) SARS-CoV-2 Main Protease Model -Absolute Quality Estimation Graph. The QMEAN Z scores were near the "0" value for the model compared to the nonredundant set of protein structures. Here the dark zone indicates that the model has a score of <1, and the model with a score of >1 or <2 or >2 locates outside of this dark zone. Good models are generally found in the dark zone. The red marker indicates a query or a generated target model, which is considered a good model according to their position near or in the dark zone (f) Structural Modelling of SARS-CoV-2 Main Protease (Swiss Model Workspace)

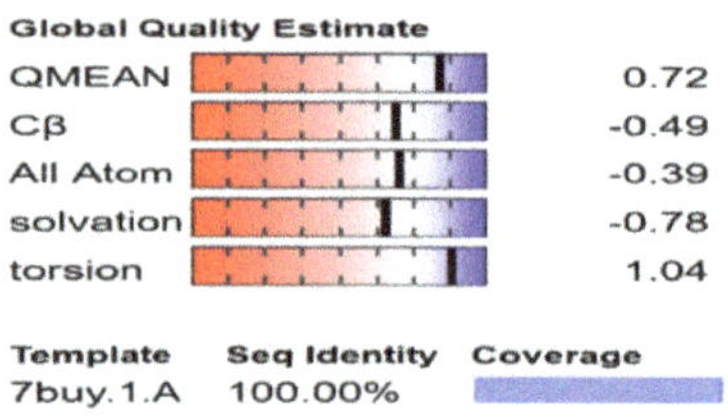

a

Scoring function terms	Z-score
QMEAN score	0.72
Cβ interaction energy	-0.49
All-atom pairwise energy	-0.39
Solvation energy	-0.78
Torsion angle energy	1.04

b

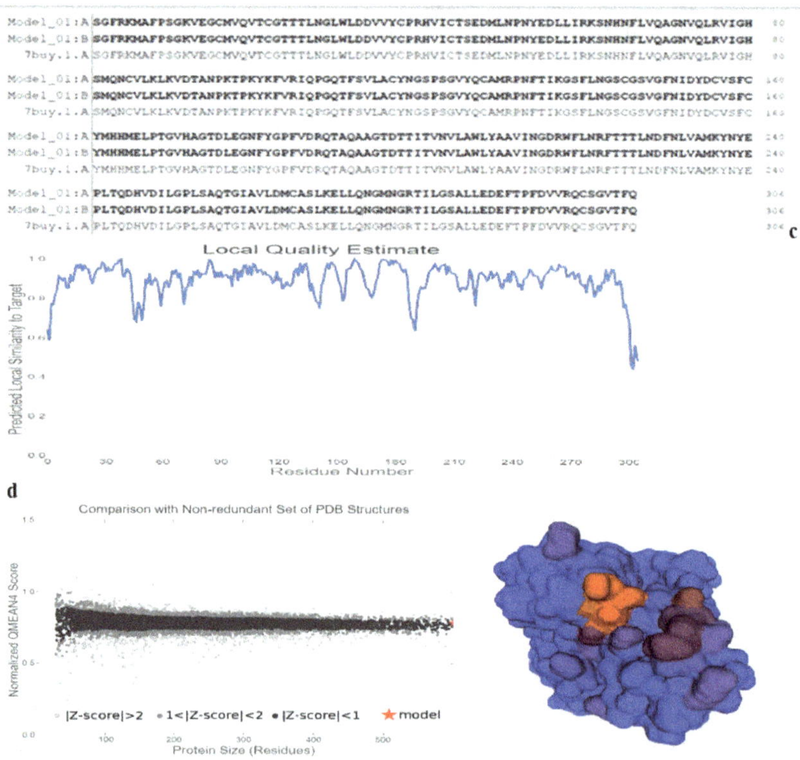

c

d

e

f

1.04 respectively for the Cβ interaction energy, all-atom pairwise energy, Solvation energy, and Torsion angle energy. Also, the model template alignment is presented in Figure 4(c). The chain A and chain B templates sequence of the COVID-19 main protease model with corresponding best (aligned) template sequence (7buy.1. A) is clearly illustrated in Figure 4(c). This is evidence that the COVID-19 main protease model has a 100% matching best template.

Furthermore, the model is shown to have a local quality estimate (LQE) with a predicted local similarity target of 1.0 in Figure 4(d). Moreover, good models are typically positioned inside the dark. The red marker in Figure 4(e) indicates a query or a generated target model, which is considered a good model according to its position near or in the dark zone. Therefore, the COVID-19 main protease model has the potential of becoming a superior model having GMQE and QMEAN scores of 0.97 and 0.72, respectively. Moreover, the model has a hundred percent aligned template, LQE prediction of local similarity target of 1.0, and the location of the model in the dark region. The surface structural model of SARS-CoV-2 Main Protease in Figure 4(f) affirms an excellent model with high quality.

The secondary structure multiple alignment unison analysis is required to establish that amino acids fall within helix strand or coil. The self-optimized prediction method with alignment (SOPMA) was used in this study to predict the structural features by advancement in secondary structure (Geourjon, & Deléage).

Table 4. Computed Secondary Structure elements of SARS-CoV-2 Main Protease by SOPMA

Secondary Structure	Element Quantity	Percentage (%)
Alpha helix (Hh)	89	29.08
3_{10} helix (Gg)	0	0
Pi helix (Ii)	0	0
Beta bridge (Bb)	0	0
Extended strand (Ee)	83	27.12
Beta turn (Tt)	35	11.44
Bend region (Ss)	0	0
Random coil (Cc)	99	32.35
Ambiguous states (?)	0	0
Other states	0	0

Parameters:
Window width: 17
Similarity threshold: 8
Number of states: 4

The predicted 3D structure shown in Table 4 reveals 29.08% Alpha helix, 27.12% Extended strand, 11.44% Beta turn, and 32.35% Random coil. The supremacy of the coiled regions affirms the high level of conservation and stability of the protein structure (Neelamathi et al.). Moreover, the side chains were re-constructed, and the regularization of the geometry of the built model was achieved with force field (Guex et al.). Figure 5 presents the default parameters for prediction as 17 window width, eight similarity threshold, and 4 number of states.

Identifying a practicable druggable binding site is the first stage of the structured-based drug design pipeline, which means identifying and delimiting protein cavities, potentially at the surface that is likely to bind to a small compound (Jiménez et al.). Proteins are made from units called amino acids, and in enzymes that are proteins, the site gets its properties from the amino acids wont to build it. These amino acids may have side chains, which will be large or small, acidic or basic, hydrophilic or hydrophobic. The set of amino acids found in the active site and their positions in 3D space give the active site a precise size, shape, and chemical behavior. An enzyme's active site is uniquely suited to bind to a particular target (the enzyme's substrate or substrates), which help them undergo a chemical reaction (https://www.khanacademy.org/science/ap-biology/cellular-energetics/enzyme-structure-and-catalysis/a/enzymes-and-the-active-site). Active site of an enzyme is that point where substrate molecule binds for the chemical reaction and generally found on the surface of the enzyme. Figure 6(d) depicts the active site of the covid-19 main

Figure 5. SARS-CoV-2 Main Protease Secondary Structure element distribution by SOPMA

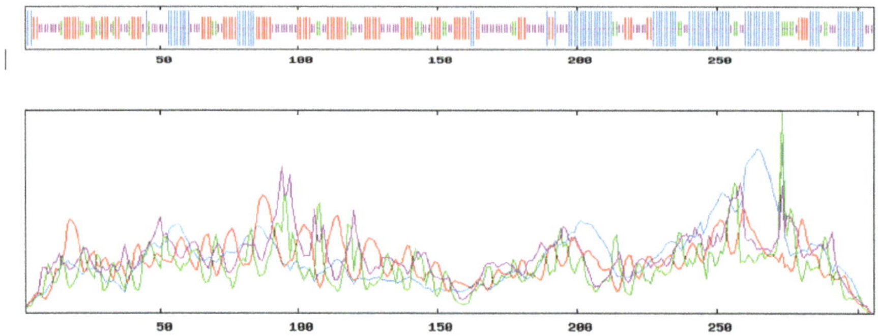

Parameters:

Window width: 17

Similarity threshold: 8

Number of states: 4

Figure 6. (a) CASTp active site chain A sequence (b) Pocket information. The table shows the area and the volume for different active sites of SARS-CoV-2 Main Protease, and the best active site remains in an area of 2451.663 and a volume of 2442.231 amino acids. (c) Atom information (d) CASTp 3Dmol.js structural visualization of SARS-CoV-2 Main Protease (e) secondary structures color-coded, highlighting residues in pockets

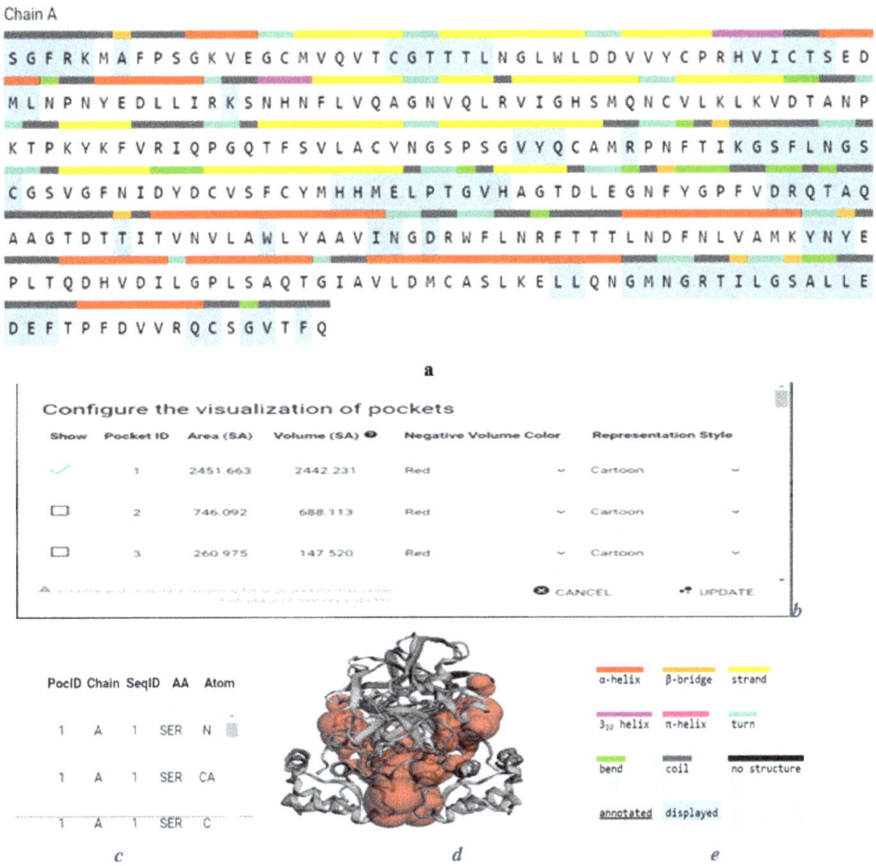

protease. The Computed Atlas of Surface Topography of proteins (CASTp) 3Dmol.js structural visualization illustrates the surface pockets, the imprint of the pocket is shown in red in Figure 6(d), and the atoms forming the pocket are shown as sticks. The CASTp 3.0server gives the best active site as an area of 2451.663, and a volume of 2442.231 amino acid is involved in the active site formation, as shown in Figure 6b. Figure 6(a) shows the active site chain A sequence, and the alphabetic letters specify different amino acids. The atom evidence is depicted in Figure 6(c), and the

Figure 7. (a) Cavity prediction by DeepSite representation by the gold-colored volume. (b) Aromatic Surface of the binding pocket of SARS-CoV-2 Main Protease. (c) Active site position of the building protein (scores and centroids)

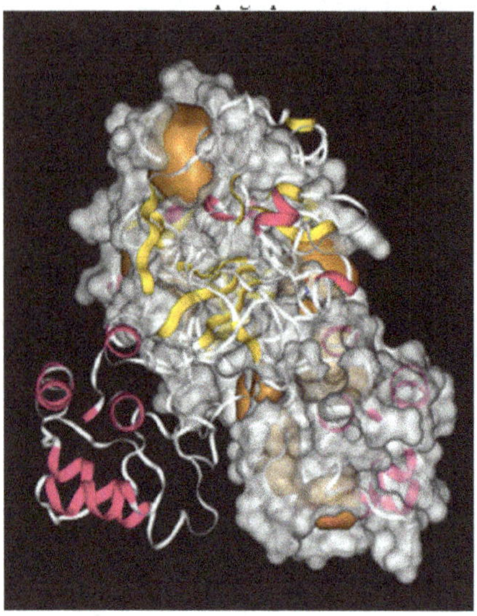

a

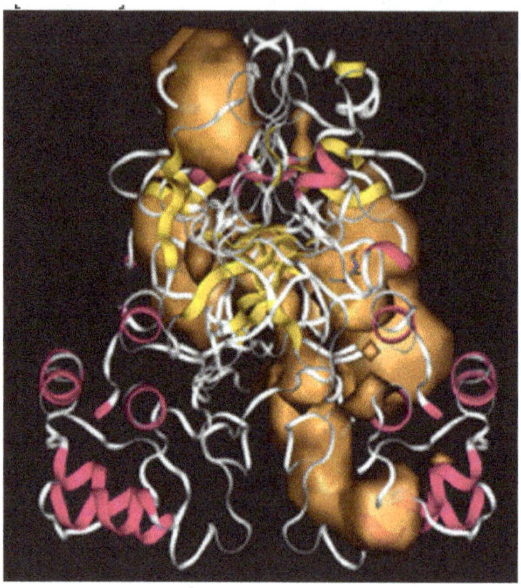

b

id	centers	score
1	[-34.8, 15.9, 52.5]	1.0
2	[-14.8, 11.9, 70.5]	1.0
3	[-14.8, 35.9, 56.5]	0.99

c

residues in pockets are shown in the secondary structures presented in Figure 6(e) (Tian et al.). The objective of the CASTp server is to provide a comprehensive and detailed quantitative characterization of topographic features of proteins (Tian et al.).

Moreover, to establish the appropriate binding site for the COVID-19 main protease, the DEEPSITE server predicted the binding pocket in the active site structure (Jiménez et al.). DeepSite is a protein binding pocket predictor based on deep neural networks (Jiménez et al.). The 3D image is notable by its grid of 1 × 1 × 1 Å3 buffer justify the pockets located to the protein edges sized voxels. The DeepSite server computed the tenancy for each property of each voxel of the 3D image of COVID-19 Main Protease, and the active site position of the building protein falls in scores and centroids shown in Figure 7(c). The gold color depicted in Figure 7(a) represents the location of the pocket. Figure 7(b) displays the binding pocket aromatic surface for the predicted COVID-19 Main Protease binding site. The pocket is shown in gold, and the atoms constituting the pocket are depicted as sticks. To confirm the drug design and discovery stage, additional experimental analysis of the protein will be necessary.

5. CONCLUSION

This study used various tools to sequence and analyze the SAR-Cov-2 (COVID-19) main protease with homology modeling to detect the binding site and reveal the classic 3D structure. The rationale for this work is providing a brace to curb the COVID-19 pandemic. This work shows that further evaluation of SAR-Cov-2 (COVID-19) main protease is essential for target drug design.

DATA AVAILABILITY

We obtained the COVID-19 target sequence from the UniProtKB (The UniProt Consortium, 2019) and extracted the experimental structure and sequence of protein template (X-ray crystallography) from the protein database (PDB).

CONFLICTS OF INTEREST

The authors declare that they have no conflicts of interest.

ACKNOWLEDGMENT

The authors would like to appreciate the support of Sol Plaatje University and Landmark University (LUCRID) for this research.

REFERENCES

Altschul, S. F., Madden, T. L., Schäffer, A. A., Zhang, J., Zhang, Z., Miller, W., & Lipman, D. J. (1997). Gapped BLAST and PSI-BLAST: A new generation of protein database search programs. *Nucleic Acids Research*, *25*(17), 3389–3402. doi:10.1093/nar/25.17.3389 PMID:9254694

Benkert, P., Biasini, M., & Schwede, T. (2010). Toward the estimation of the absolute quality of individual protein structure models. *Bioinformatics (Oxford, England)*, *27*, 343–345.

Arulmozhiyal & Baskaran. (n.d.). Implementation of a Fuzzy PI Controller for Speed Control of Induction Motors Using FPGA. *Journal of Power Electronics*, *10*, 65–71.

Bertoni, M., Kiefer, F., Biasini, M., Bordoli, L., & Schwede, T. (2017). Modeling protein quaternary structure of homo- and hetero-oligomers beyond binary interactions by homology. *Scientific Reports*, *7*(1), 7. doi:10.103841598-017-09654-8 PMID:28874689

Bienert, S., Waterhouse, A., de Beer, T. A. P., Tauriello, G., Studer, G., Bordoli, L., & Schwede, T. (2017). The SWISS-MODELRepository - new features and functionality. *Nucleic Acids Research*, *45*(D1), D313–D319. doi:10.1093/nar/gkw1132 PMID:27899672

Bowie, J. U., Lüthy, R., & Eisenberg, D. (1991, July 12). A method to identify protein sequences that fold into a known three-dimensional structure. *Science*, *253*(5016), 164–170. doi:10.1126cience.1853201 PMID:1853201

Camacho, C., Coulouris, G., Avagyan, V., Ma, N., Papadopoulos, J., Bealer, K., & Madden, T. L. (2009). BLAST+: Architecture and applications. *BMC Bioinformatics*, *10*(1), 421–430. doi:10.1186/1471-2105-10-421 PMID:20003500

Chellapandi, P., & Saranya, S. (2020). Genomics insights of SARS-CoV-2 (COVID-19) into target-based drug discovery. *Medicinal Chemistry Research*, 1–15. doi:10.1007/s00044-020-02610-8

Cheng, J., Randall, A. Z., Sweredoski, M. J., & Baldi, P. (2005). SCRATCH: A protein structure and structural feature prediction server. *Nucleic Acids Research*, *33*(Web Server), 72–76. doi:10.1093/nar/gki396 PMID:15980571

Combet C., Blanchet C., Geourjon C., & Deléage G. (2000). NPS@: Network Protein Sequence Analysis. *TIBS, 25*(3), 147-150

Fox, D. (2020). What you need to know about the Wuhan coronavirus. *Nature*. doi:10.1038/d41586-020-00209-y PMID:33483684

Geourjon, C., & Deléage, G. (1995). SOPMA: Significant improvement in protein secondary structure prediction by consensus prediction from multiple alignments. *Cabios*, *11*(6), 681–684. doi:10.1093/bioinformatics/11.6.681 PMID:8808585

Guenther, S., Reinke, P., Oberthuer, D., Yefanov, O., Gelisio, L., Ginn, H., Lieske, J., Domaracky, M., Brehm,W., Rahmani Mashour, A., White, T.A., Knoska, J., Pena Esperanza, G., Koua, F., Tolstikova, A., Groessler, M., Fischer, P., Hennicke, V., Fleckenstein, H., Trost, F., … Wolle, J. (2020). Structure of SARS-CoV-2 Main Protease bound to 2-Methyl-1-tetralone. *Protein Data Bank*. doi:10.2210/pdb6ynq/pdb

Guex, N., Peitsch, M. C., & Schwede, T. (2009). Automated comparative protein structure modeling with SWISS-MODEL and swiss-PdbViewer: A historical perspective. *Electrophoresis*, *30*(S1), S162–S173. doi:10.1002/elps.200900140 PMID:19517507

Jiménez, J., Doerr, S., Martínez-Rosell, G., Rose, A. S., & De Fabritiis, G. (2017). DeepSite: Protein-binding site predictor using 3D-convolutional neural networks. *Bioinformatics (Oxford, England)*, *33*(19), 3036–3042. doi:10.1093/bioinformatics/btx350 PMID:28575181

Laskowski, R. A., Jabłońska, J., Pravda, L., Vařeková, R. S., & Thornton, J. M. (2018). PDBsum: Structural summaries of PDB entries. *Protein Science*, *27*(1), 129–134. doi:10.1002/pro.3289 PMID:28875543

Laskowski, R. A., Rullmannn, J. A., MacArthur, M. W., Kaptein, R., & Thornton, J. M. (1996). AQUA and PROCHECK-NMR: Programs for checking the quality of protein structures solved by NMR. *Journal of Biomolecular NMR*, *8*(4), 477–486. doi:10.1007/BF00228148 PMID:9008363

Lavecchia, A., & Giovanni, C. (n.d.). Virtual Screening Strategies in Drug Discovery: A Critical Review. *Current Medicinal Chemistry*, *20*, 2839–2860.

Lüthy, R., Bowie, J. U., & Eisenberg, D. (1992, March 5). Assessment of protein models with three-dimensional profiles. *Nature*, *356*(6364), 83–85. doi:10.1038/356083a0 PMID:1538787

Mirdita, M., von den Driesch, L., Galiez, C., Martin, M. J., Söding, J., & Steinegger, M. (2009). Uniclust databases of clustered and deeply annotated protein sequences and alignments. *Nucleic Acids Research*, *45*, D170–D176.

Neelamathi, E., Vasumathi, E., Bagyalakshmi, S., & Kannan, R. (2016). Insilico prediction of structure and functional aspects of a hypothetical protein of Neurospora crassa. *J Cell Tissue Re*, *9*, 1889–1894.

Nielsen, M., Lundegaard, C., Lund, O., & Petersen, T. N. (2010). CPHmodels-3.0 - Remote homology modeling using structure-guided sequence profiles. *Nucleic Acids Research*, *38*(suppl_2), W576–W581. doi:10.1093/nar/gkq535 PMID:20542909

Premalatha, D., Ravindra, P., & Venkateshwer Rao, L. (2007). Homology modeling of putative thioredoxin from Helicobacerr pylori. *Indian Journal of Biotechnology*, *6*, 4859.

Ripphausen, P., Nisius, B., Peltason, L., & Bajorath, J. (2010). Quo Vadis, Virtual Screening? A Comprehensive Survey of Prospective Applications. *Journal of Medicinal Chemistry*, *53*(24), 8461–8467. doi:10.1021/jm101020z PMID:20929257

Schaller, D., Hagenow, S., Stark, H., & Wolber, G. (2019). Ligand-guided homology modeling drives identification of novel histamine H3 receptor ligands. *PLoS One*, *14*(6), e0218820. doi:10.1371/journal.pone.0218820 PMID:31237914

Steinegger, M., Meier, M., Mirdita, M., Vöhringer, H., Haunsberger, S. J., & Söding, J. (2019). HH-suite3 for fast remote homology detection and deep protein annotation. *BMC Bioinformatics*, *20*(1), 473. doi:10.118612859-019-3019-7 PMID:31521110

Studer, G., Rempfer, C., Waterhouse, A. M., Gumienny, G., Haas, J., & Schwede, T. (2020). QMEANDisCo - distance constraints applied on model quality estimation. *Bioinformatics (Oxford, England)*, *36*(6), 1765–1771. doi:10.1093/bioinformatics/btz828 PMID:31697312

The UniProt Consortium. (2019). UniProt: A worldwide hub of protein knowledge. *Nucleic Acids Research*, *47*(D1), D506–D515. doi:10.1093/nar/gky1049 PMID:30395287

Tian, W., Chen, C., Lei, X., Zhao, J., & Liang, J. (2018, July 2). CASTp 3.0: Computed atlas of surface topography of proteins. *Nucleic Acids Research*, *46*(W1), W363–W367. doi:10.1093/nar/gky473 PMID:29860391

Waterhouse, A., Bertoni, M., Bienert, S., Studer, G., Tauriello, G., Gumienny, R., Heer, F. T., de Beer, T. A. P., Rempfer, C., Bordoli, L., Lepore, R., & Schwede, T. (2013). SWISS-MODEL: homology modeling of protein structures and complexes. Nucleic Acids Res., 46(W1), W296-W303.

Chapter 5
Convolutional Neural Networks (CNN): Models and Algorithms

Friki Imed Salaheddine
Dr. Moulay Tahar University, Algeria

ABSTRACT

Deep learning is one of the main technologies of machine learning. With deep learning, this chapter is talking about algorithms capable of mimicking the actions of the human brain through artificial neural networks. Compared to other algorithmic structures, neural networks have great advantages: first, their structure based on the stacking of non-linear functions gives them an enormous capacity. In this chapter, the authors will encompass the different models of the convolutional neural network that exist in the literature, their parameters, their architectures, their layers, their advantages, and their disadvantages. The objective is to help researchers to have a summary on deep learning.

INTRODUCTION

Deep learning is an ambiguous term, as it has gone through several different meanings throughout the years.

Deep learning is a form of machine learning that enables computers to learn from experience and understand the world in terms of a hierarchy of concepts. Because the computer gathers knowledge from experience, there is no need for a human computer operator formally to specify all of the knowledge needed by the computer.

DOI: 10.4018/978-1-6684-5656-9.ch005

Deep Learning is a subset of machine learning, which is in itself a subfield of AI. The figure below is a visual representation of the relationship between AI, ML and DL.

Deep learning, with its remarkable ability to derive meaning from complicated or imprecise data, can be used to extract patterns and detect trends too complex to be noticed by humans or computer techniques. A trained neural network can be considered an "expert" in the category of information it has been given to analyse. This expert can then be used to provide new projections of situations of interest and response. Other benefits include

1. Storage of information on the whole network: information such as in traditional programming is stored on the whole network, not on a database.
2. Ability to work with incomplete knowledge: after training, the data can produce an output even with incomplete information.
3. Fault tolerance: the corruption of one or more cells.
4. Having a distributed memory: for Ann to be able to learn, it is necessary to determine the instances and teach the network according to the desired output by showing these instances to the network. The success of the network is directly proportional to the instances selected and if the event cannot be shown to the network in all its aspects, the network may produce a false output.
5. Ability to make machine learn: Artificial neural networks learn from events and make decisions by commenting on similar events.
6. Parallel processing capability: artificial neural networks have a numerical strength that can perform more than one job at the same time (Jakhar, 2020)

The rest of the chapter is structured as following: section two we will present the convolutional neural network family. Section three we will detail the parameters like padding, pooling activation function and edge detection techniques. Section four, illustrates classification models like VGG16, vgg19, AlexNet, GoogleNet ...ect. Section 5 describes the segmentation convolutional model as UNet. Finally, we conclude with a general conclusion and some future works.

CONVOLUTIONAL NEURAL NETWORKS

In recent years, the convolutional neural network (CNN) has emerged as the prevalent model for the machine learning and computer vision. Deep CNNs are widely used in a broad range of real-life applications, such as image classification, object detection and image segmentation. The name "convolutional neural network" indicates that the network employs a mathematical operation called convolution. Convolution is

Figure 1. Relationship between AI, ML and DL (Jakhar, 2020)

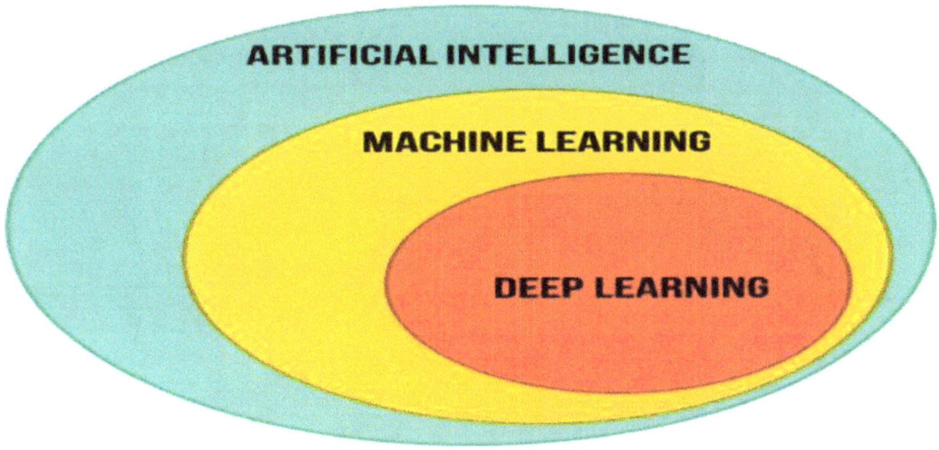

a specialized kind of linear operation. Convolutional networks are simply neural networks that use convolution in place of general matrix multiplication in at least one of their layers. The typical use of convolutional networks is on classification tasks, where the output to an image is a single class label.

Overall Architecture

CNNs are comprised of three types of layers. These are convolutional layers, pooling layers and fully-connected layers. When these layers are stacked, A simplified CNN architecture for MNIST classification is illustrated in Figure 2 (Lee, 2018).

The basic functionality of the example CNN above can be broken down into four key areas.

a. As found in other forms of ANN, the input layer will hold the pixel values of the image.
b. The convolutional layer will determine the output of neurons of which are connected to local regions of the input through the calculation of the scalar product between their weights and the region connected to the input volume. The rectified linear unit (commonly shortened to ReLu) aims to apply an 'elementwise' activation function such as sigmoid to the output of the activation produced by the previous layer.
c. The pooling layer will then simply perform downsampling along the spatial dimensionality of the given input, further reducing the number of parameters within that activation.

Figure 2. General Architecture of Convolutional neural network algorithms

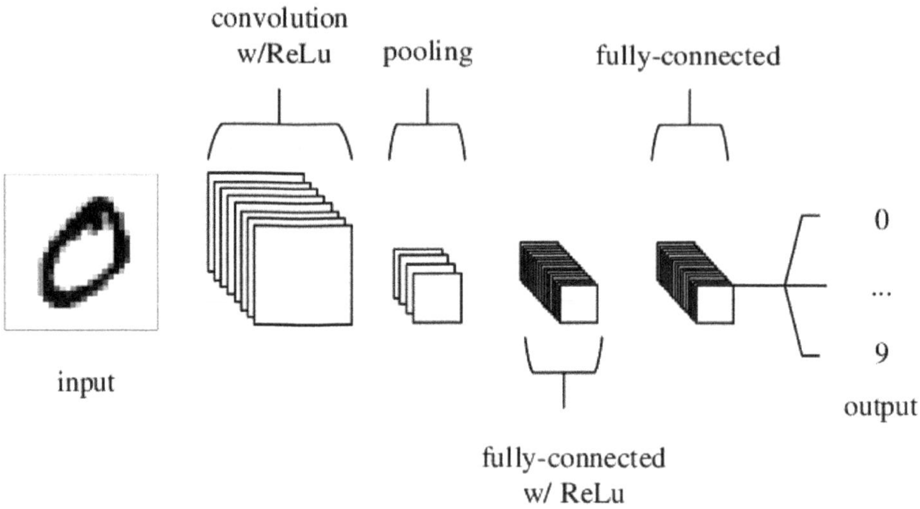

d. The fully-connected **layers** will then perform the same duties found in standard ANNs and attempt to produce class scores from the activations, to be used for classification. It is also suggested that ReLu may be used between these layers, as to improve performance (Basha, 2020).

Edge Detection

The edge detection on the images is so important for image processing. It is used in a various fields of applications ranging from real-time video surveillance and traffic management to medical imaging applications. The role of edge detection is very crucial, as it is the preliminary or fundamental stage in pattern recognition. Edges characterize object boundaries and are therefore useful for segmentation and identification of objects in a scene. The idea that the edge detection is the first step in vision processing has fueled a long-term search for a good edge detection algorithm (Jing, 2022).

Padding

Convolutional operation often requires padding when part of the filter extends beyond the input image or feature map. Standard approaches include zero padding (extend with zeros), reflection padding (reflect the input values across the border axis) and replication padding (extend by replicating the values along borders) padding (extend

Figure 3. Edge detection using Sobel filters (Jing, 2022)

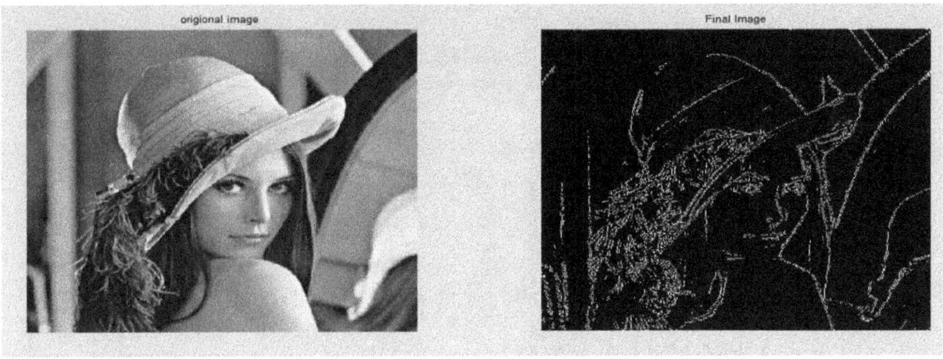

by replicating the values along borders). Among them, the most commonly used scheme is zero padding, as was adopted by.

Padding basically extends the area of an image in which a convolutional neural network processes. The kernel/filter which moves across the image scans each pixel and converts the image into a smaller image. In order to work the kernel with processing in the image, padding is added to the outer frame of the image to allow for more space for the filter to cover in the image. Adding padding to an image processed by a CNN allows for a more accurate analysis of images.

Types of Padding

1. The Valid Convolutions: "No padding", its mean

N.N ∗ F.F −® (N − F +1)∗(N − F +1) (1)

2. The Same convolutions: Dimension of input matrix = dimension of output matrix

(N.N)∗(F.F) −® ((N +2P − F)/S)+1∗((N +2P − F)/S)+1 (2)

Through this equation we find this rule to achieve the same convolutions:

P = (F −1)/2 (3)

3. Strided convolutions It's helpful for save time but the quality decreases

(N.N)∗(F.F) =Þ ((N +2P − F)/S)+1∗((N +2P − F)/S)+1 (4)

Pooling

Pooling layers aim to gradually reduce the dimensionality of the representation, and thus further reduce the number of parameters and the computational complexity of the model.

Figure 4. ConvNet application using a single padding (Liu, 2022)

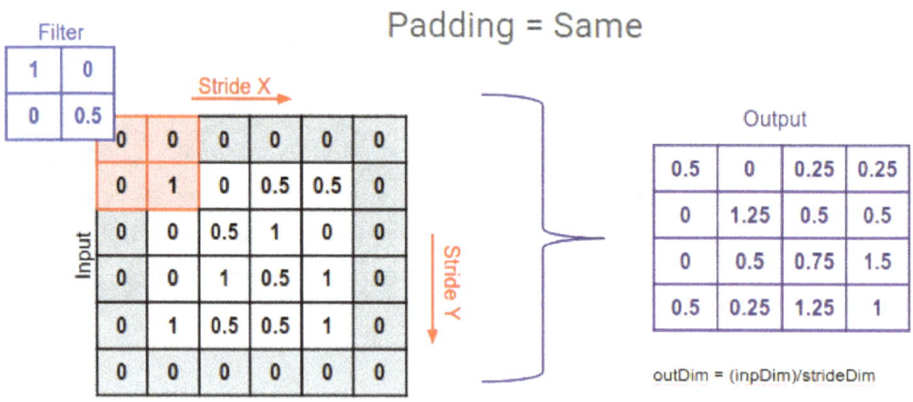

The pooling layer operates over each activation map in the input, and scales its dimensionality using the "MAX" function. In most CNNs, these come in the form of max-pooling layers with kernels of a dimensionality of 2 × 2 applied with a stride of 2 along the spatial dimensions of the input. This scales the activation map down to 25% of the original size - whilst maintaining the depth volume to its standard size.

Due to the destructive nature of the pooling layer, there are only two generally observed methods of max-pooling. Usually, the stride and filters of the pooling layers are both set to 2 × 2, which will allow the layer to extend through the entirety of the spatial dimensionality of the input. Furthermore overlapping pooling may be utilised, where the stride is set to 2 with a kernel size set to 3. Due to the destructive nature of pooling, having a kernel size above 3 will usually greatly decrease the performance of the model.

It is also important to understand that beyond max-pooling, CNN architectures may contain general-pooling. General pooling layers are comprised of pooling neurons that are able to perform a multitude of common operations including L1/ L2-normalisation, and average pooling. However, this tutorial will primarily focus on the use of max-pooling.

Activation Functions

Activation Functions are specially used in artificial neural networks to transform an input signal into an output signal which in turn is fed as input to the next layer in the stack. In an artificial neural network, we calculate the sum of products of inputs and their corresponding weights and finally apply an activation function to

Figure 5. Activation function on a single neuron (Nwankpa, 2018)

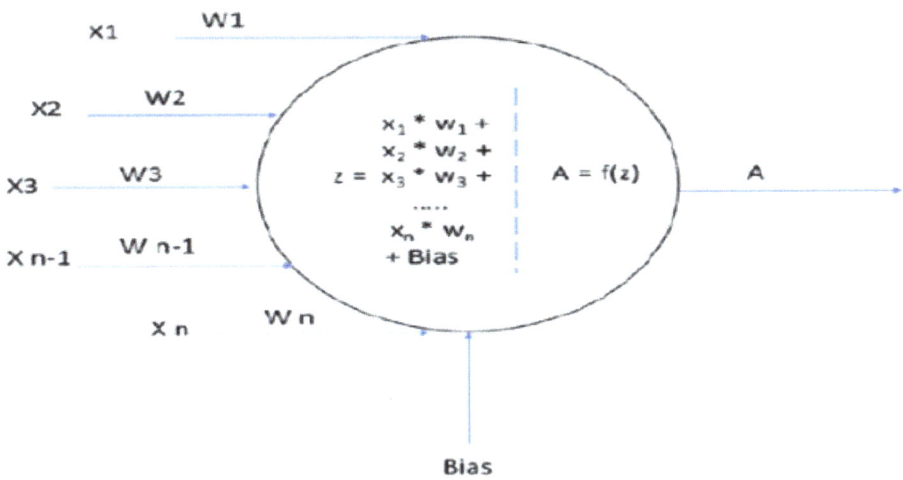

it to get the output of that particular layer and supply it as the input to the next layer (Nwankpa, 2018).

Some examples of some of the popular activation functions used in neural networks for classification and detection tasks are given as follows:

Sigmoid Activation Function

It is the most widely used activation function as it is a non-linear function. Sigmoid function transforms the values in the range 0 to 1. It can be defined as:

f(x) = 1/e-x

Sigmoid function is continuously differentiable and a smooth S-shaped function. The derivative of the function is:

f'(x) = 1-sigmoid(x)

Also, sigmoid function is not symmetric about zero which means that the signs of all output values of neurons will be same. This issue can be improved by scaling the sigmoid function.

Tanh Function:

It is Hyperbolic Tangent function. Tanh function is similar to the sigmoid function but it is symmetric to around the origin. This results in different signs of outputs from previous layers which will be fed as input to the next layer. It can be defined as:

f(x) = 2sigmoid(2x)-1

Figure 6. Sigmoid function (Nwankpa, 2018)

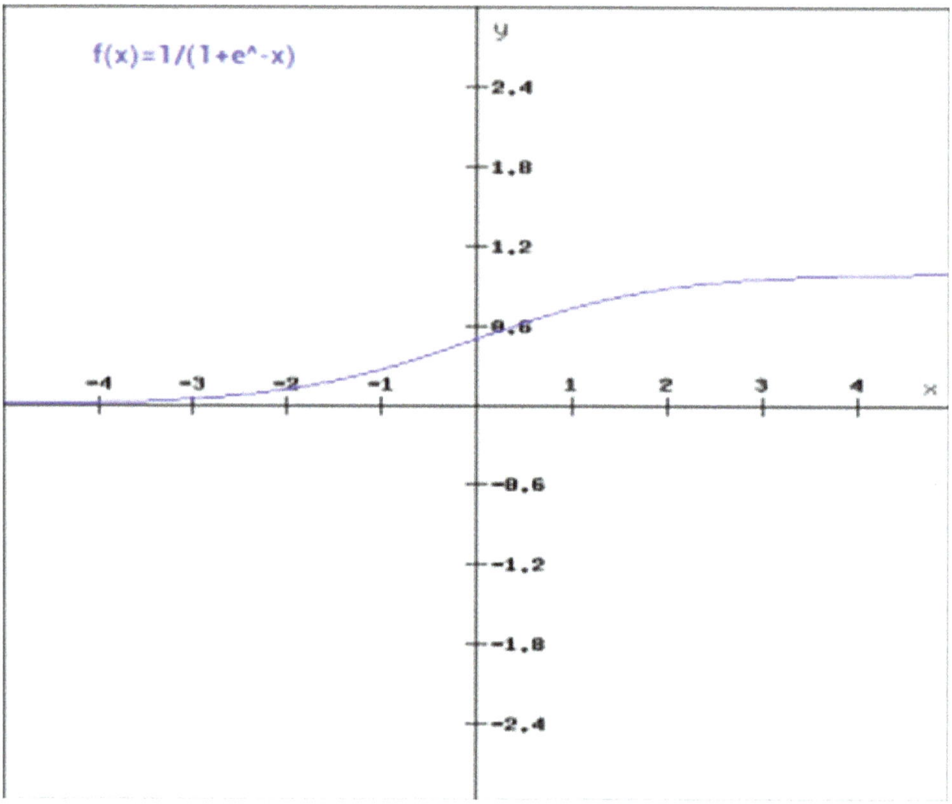

Tanh function is continuous and differentiable, the values lies in the range -1 to 1. As compared to the sigmoid function the gradient of tanh function is more steep. Tanh is preferred over sigmoid function as it has gradients which are not restricted to vary in a certain direction and also, it is zero centered. (Sharma, 2017)

ReLU Function

ReLU stands for rectified liner unit and is a non-linear activation function which is widely used in neural network.

The upper hand of using ReLU function is that all the neurons are not activated at the same time. This implies that a neuron will be deactivated only when the output of linear transformation is zero. It can be defined mathematically as:

$f(x) = \max(0, x)$

ReLU is more efficient than other functions because as all the neurons are not activated at the same time, rather a certain number of neurons are activated at a time.

Figure 7. Tanh Function (Nwankpa, 2018)

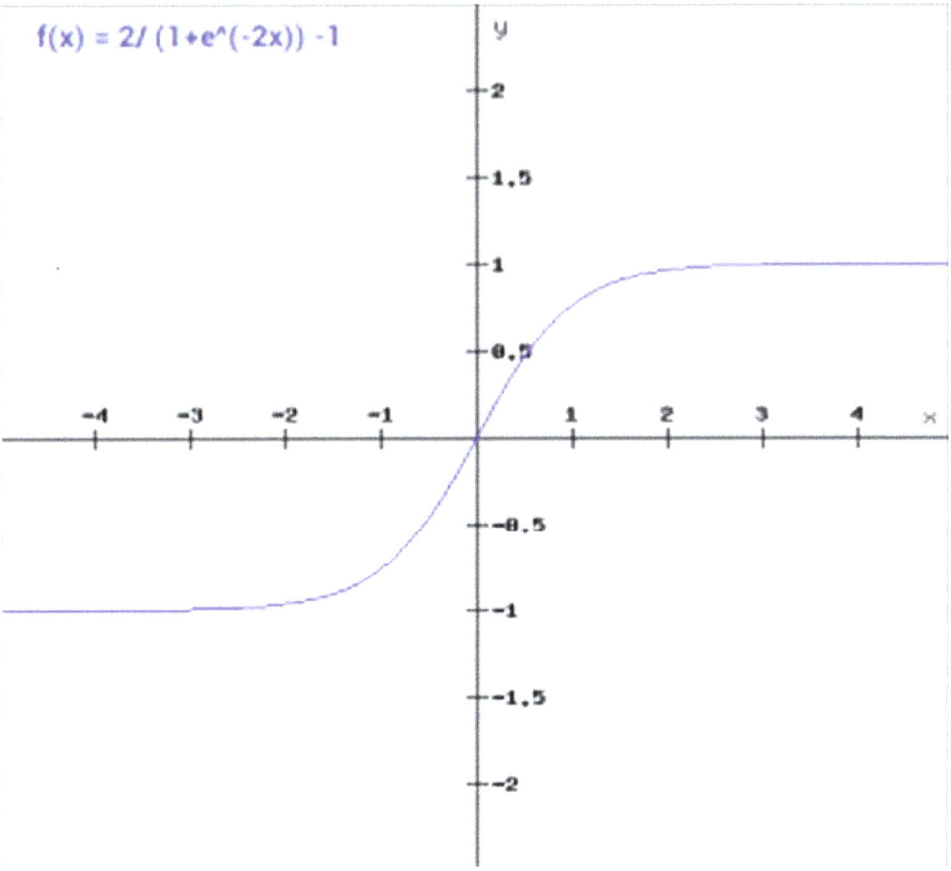

In some cases, the value of gradient is zero, due to which the weights and biases are not updated during back-propagation step in neural network training.

Leaky ReLU

The Leaky ReLU activation function is similar to the ReLU function with one key difference: it allows a small positive gradient when the unit is inactive. A Leaky ReLU is calculated as follows:

(x)=max $(x,)$. (5)

Figure 8. ReLU Function (Nwankpa, 2018)

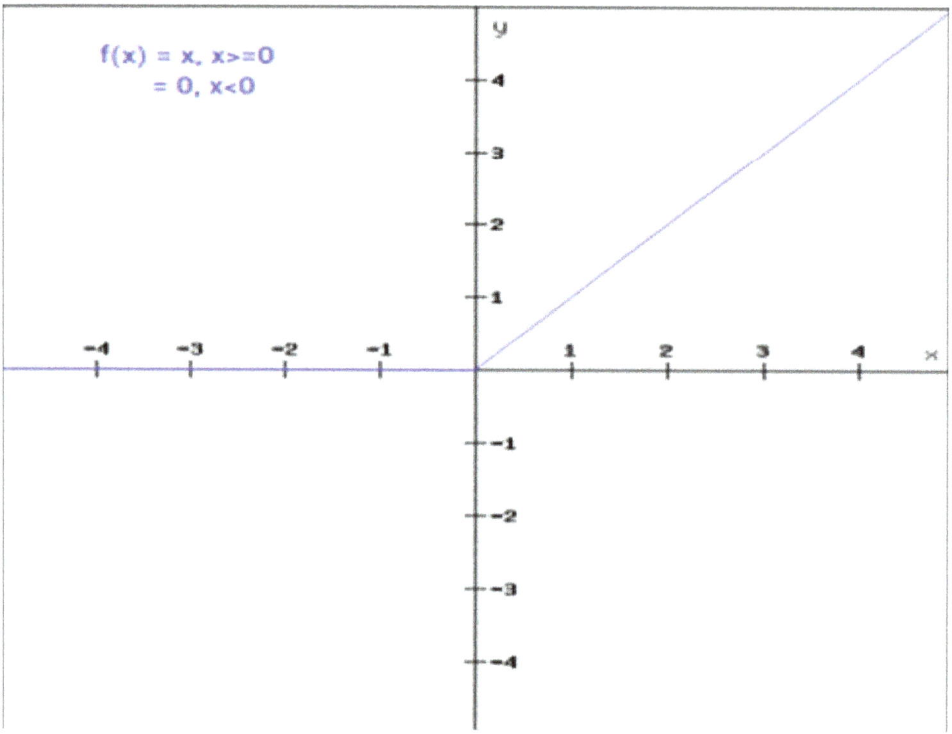

Softmax Activation Function:

Softmax function is a combination of multiple sigmoid functions. As we know that a sigmoid function returns values in the range 0 to 1, these can be treated as probabilities of a particular class' data point.

Softmax function unlike sigmoid functions which are used for binary classification, can be used for multiclass classification problems. The function, for every data point of all the individual classes, returns the probability. It can be expressed as:

$$\sigma(z)_j = \frac{e^{z_j}}{\sum_{K=1}^{K} e^{z_k}} \text{ for j} = 1, \ldots, \text{K}. \tag{6}$$

When we build a network or model for multiple class classification, then the output layer of the network will have the same number of neurons as the number of classes in the target (Nwankpa, 2018).

Figure 9. Leaky ReLU Function (Nwankpa, 2018)

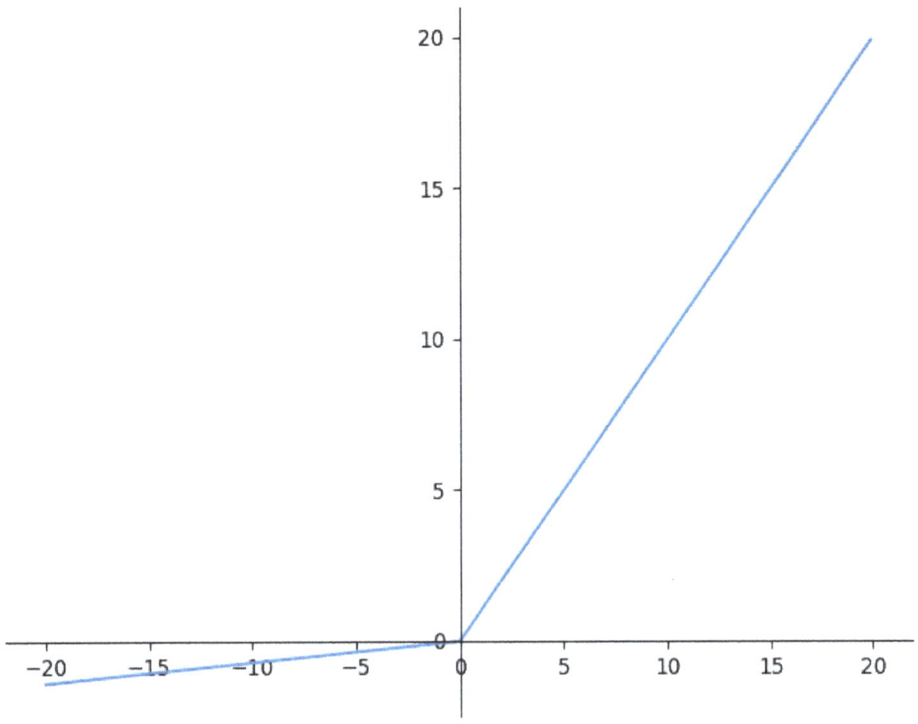

Optimization Methods

The most important part of the model training is the optimizer. Let's imagine the structure of a model. The structure created by defining the sequence of layers with the number of neurons, activation functions and the shape of input and output is initialized with random weights at the beginning. The weights that determined the influence of a neuron on the next neuron or the final output are updated by the network during the learning process. In short, a network with random weights and a defined structure is the starting point for a model. The network takes a training sample and uses its values as inputs to the neurons in the first layer, which then produce an output with the defined activation function. The output then becomes an input to the next layer, and so on. The output of the last layer would be the prediction for the training sample. This is where the loss function comes into play. The loss function helps the network understand how well or poorly the current set of weights performed on the training sample. The next step for the model is to reduce the loss. How does the network know what steps or updates it needs to make to the weights to

Figure 10. The loss score is used as a feedback signal to adjust the weights (Nwankpa, 2018)

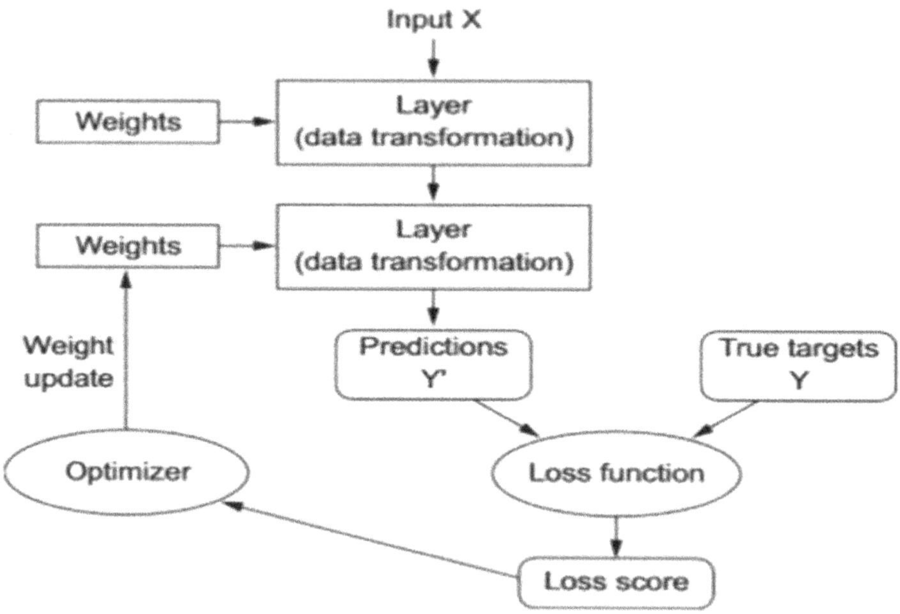

reduce the loss? The optimizer helps it understand this step. The optimizer function is a mathematical algorithm that uses derivatives, partial derivatives, and the chain rule in the calculation to understand how much the network will change in the loss function by making a small change to the weights of the neurons. The change in the loss function, which would be an increase or decrease, helps to determine the direction of change needed in the weight of the connection.

To summarize, we can say that the optimizer uses the score obtained by the loss function to adjust the value of the weights a bit, in a direction that will lower the loss score, as shown in Figure bellow. The optimizer implements what is called the backpropagation algorithm: the central algorithm of deep learning

Stochastic Gradient Descent (SGD)

SGD performs an iteration with each training sample (i.e., after the pass of every training sample, it calculates the loss and updates the weight). Since the weights are updated too frequently, the overall loss curve would be very noisy. However, the optimization is relatively fast compared to others.

The formula for weight updates can be expressed in a simple way as follows:

Weights = Weights – learning rate * Loss (7)

Where learning rate is a parameter we define in the network architecture. Say, for learning rate =0.01.

For updates with every training sample, we would need to use batch_ size=1 in the model training function. To reduce high fluctuations in the SGD optimizations, a better approach would be to reduce the number of iterations by providing a minibatch, which would then enable averaging the loss for all samples in a batch and updating the weights at the end of the batch. This approach has been more successful and results in a smoother training process. Batch size is usually set in powers of 2 (i.e., 32, 64, 128, etc.) (Duan, 2021).

Adam

Adam, which stands for adaptive moment estimation, is by far the most popular and widely used optimizer in DL. In most cases, you can blindly choose the Adam optimizer and forget about the optimization alternatives. This optimization technique computes an adaptive learning rate for each parameter. It defines momentum and variance of the gradient of the loss and leverages a combined effect to update the weight parameters. The momentum and variance together help smooth the learning curve and effectively improve the learning process. (Gulli, 2017). The math representation can be simplified in the following way:

Weights = Weights – (Momentum and Variance combined) (8)

Other Important Optimizers

There are many other popular optimizers that can also be used for different DL models like Adagrad, Adadelta, RMSProp, Adamax and Nadam. Each of the optimization techniques has its own pros and cons. A major problem which we often face in DL is the vanishing gradient and saddle point problem. You can explore these problems in more detail while choosing the best optimizer for your problem. But for most use cases, Adam always works fine.

Architectural Innovations in CNN

Different improvements in CNN architecture have been made from 1989 to date. These improvements can be categorized as parameter optimization, regularization, structural reformulation, etc. However, it is observed that the main thrust in CNN performance improvement came from the restructuring of processing units and the designing of new blocks. Most of the innovations in CNN architectures have been made in relation to depth and spatial exploitation. Depending upon the

Figure 11. Taxonomy of deep CNN architectures showing seven different categories (Duan, 2021)

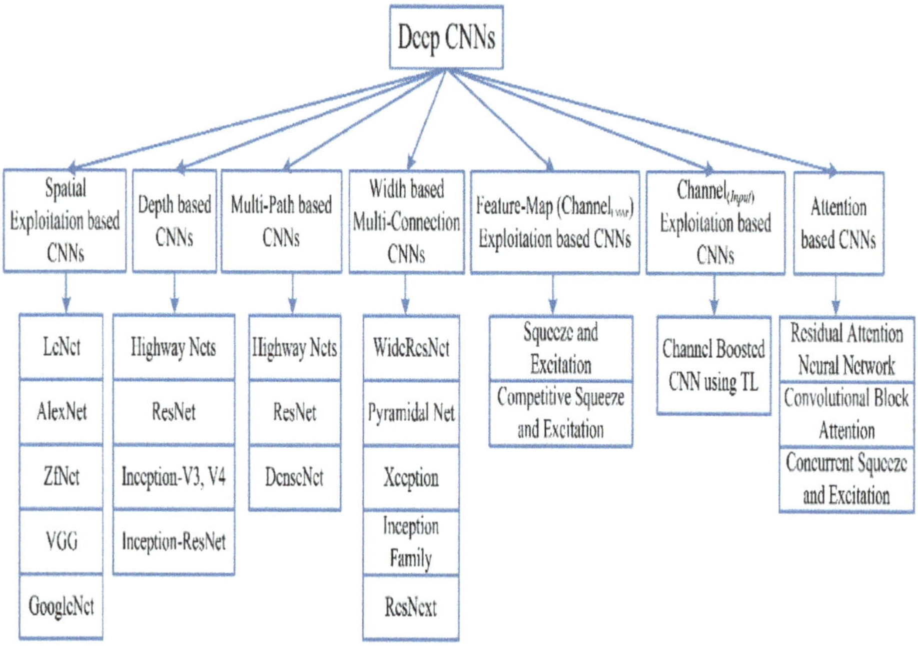

type of architectural modifications, CNNs can be broadly categorized into seven different classes, namely; spatial exploitation, depth, multi-path, width, feature-map exploitation, channel boosting, and attention-based CNNs. The taxonomy of CNN architectures is pictorially represented in Fig. 12. Architectural details of the state-of-the-art CNN models, their parameters, and performance on benchmark datasets are summarized in Table 1 (Khan, 2020).

DIFFERENT MODELS OF CNN FOR IMAGES CLASSIFICATION

LeNet-5

LeNet-5 is a gradient-based learning CNN structure and first successfully applied in hand-written digital character recognition (Wei, 2019), introduced by Yan LeCun as shown in figure bellow, has 7 weighted (trainable) layers. Among them, three (C1, C3, C5) convolutional layers, two (S2, S4) average pooling layers, one (F6) fully connected layer and one output layer (Sultana, 2018).

Figure 12. Architecture of Lenet-5 (Duan, 2021)

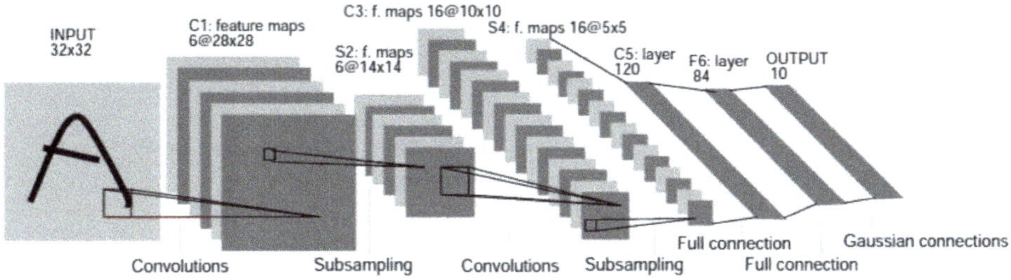

AlexNet

In 2012 Krizhevky et al. designed a large deep CNN, called AlexNet to classify ImageNet (Alom, 2018) data. The architecture of AlexNet is same as LeNet-5 but much bigger. It is made up of 8 trainable layers. Among them, 5 convolutional layers (conv layer) and 3 fully connected layers are there. Using rectified linear unit (ReLU) (Duan, 2021).

Figure 13. Architecture of AlexNet (Alom, 2018)

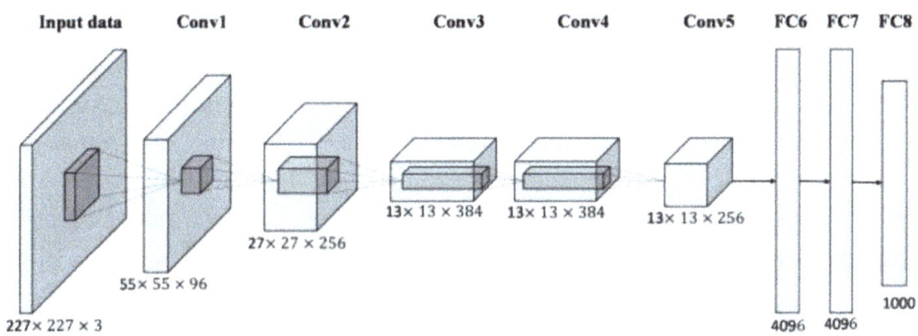

VGGNet

The Visual Geometry Group network (VGGNet) is a deep neural network with a multilayered operation. The VGGNet is based on the CNN model and is applied on the ImageNet dataset (Alom, 2018).

Figure 14. VGG-16 CNN model architecture layer wise (Srivastava, 2020)

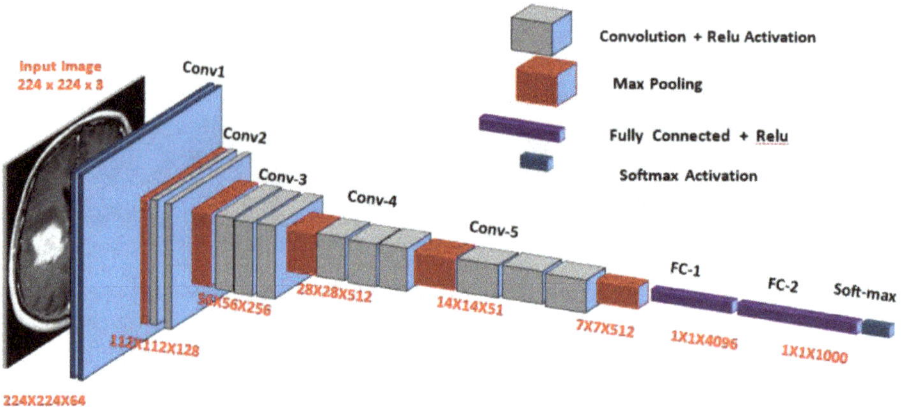

Simonyan and Zisserman used deeper configuration of AlexNet, and they proposed it as VGGNet. They have used small filters of size 3×3 for all layers and made the network deeper keeping other parameters fixed.

VGG16

VGG-16 model is one of the simple architecture of CNN which uses less hyperparameter. This model uses filter size 3×3 throughout the architecture and also it uses 1 stride in the convolution layer and stride 2 for the pooling layer with the same padding. It consists of 16 layers including the convolution layer to softmax that is why it is named VGG-16. The last layer of the VGG16 model is replaced by two FC layer with softmax activation instead of 10 FC layer with softmax (Alom, 2018). And the detailed layer by layer configuration of the VGG-16 model is pictured in figure bellow.

VGG19

Has 19 weight layers consisting of 16 convolutional layers with 3 fully connected layers and same 5 pooling layers. In both variation of VGGNet there consists of two Fully Connected layers with 4096 channels each which is followed by another fully connected layer with 1000 channels to predict 1000 labels. Last fully connected layer uses softmax layer for classification purpose.

Residual Network (ResNet)

The winner of ILSVRC 2015 was the Residual Network architecture, ResNet. Resnet was developed by Kaiming He with the intent of designing ultra-deep networks that did not suffer from the vanishing gradient problem that predecessors had.

ResNet is developed with many different numbers of layers; 34, 50,101, 152, and even 1202. The popular ResNet50 contained 49 convolution layers and 1 fully connected layer at the end of the network. The total number of weights and MACs for the whole network are 25.5M and 3.9M respectively.

The basic block diagram of the ResNet architecture is shown in Fig. 18. ResNet is a traditional feed forward network with a residual connection. The output of a residual layer can be defined based on the outputs of $(l - 1)$ *th* which comes from the previous layer defined as $xl-1$. $\mathcal{F}$ $(xl-1)$ is the output after performing various operations (e.g. convolution with different size of filters, Batch Normalization (BN) followed by an activation function such as a ReLU on $xl-1$). The final output of residual unit which can be defined with the following equation:

xl = F(xl-1)+ xl-1 (8)

The residual network consists of several basic residual blocks. However, the operations in the residual block can be varied depending on the different architecture of residual networks. The wider version of residual network was proposed by Zagoruvko el at. In 2016. Another improved residual network approach known as aggregated residual transformation was proposed in 2016. Recently, some other variants of residual models have been proposed based on the Residual Network architecture. Furthermore, there are several advanced architectures that have been proposed with the combination of Inception and Residual units. The basic conceptual diagram of Inception-Residual unit is shown in the following Figure 15.

Mathematically, this concept can be represented as

$$x_l = F(x_{l-1}^{3\times3} \odot x_{l-1}^{5\times5}) + x_{l-1} \quad (9)$$

Where the symbol $\odot$ refers the concentration operations between two outputs from the 3×3 and 5×5 filters. After that, the convolution operation is performed with1×1 filters.

Finally, the outputs are added with the inputs of this block of xl-1. The concept of Inception block with residual connections is introduced in the Inception-v4 architecture. The improved version of the Inception-Residual network was also proposed (Alom, 2018).

Figure 15. Basic diagram of residual block (Abdurakhmonov, 2016)

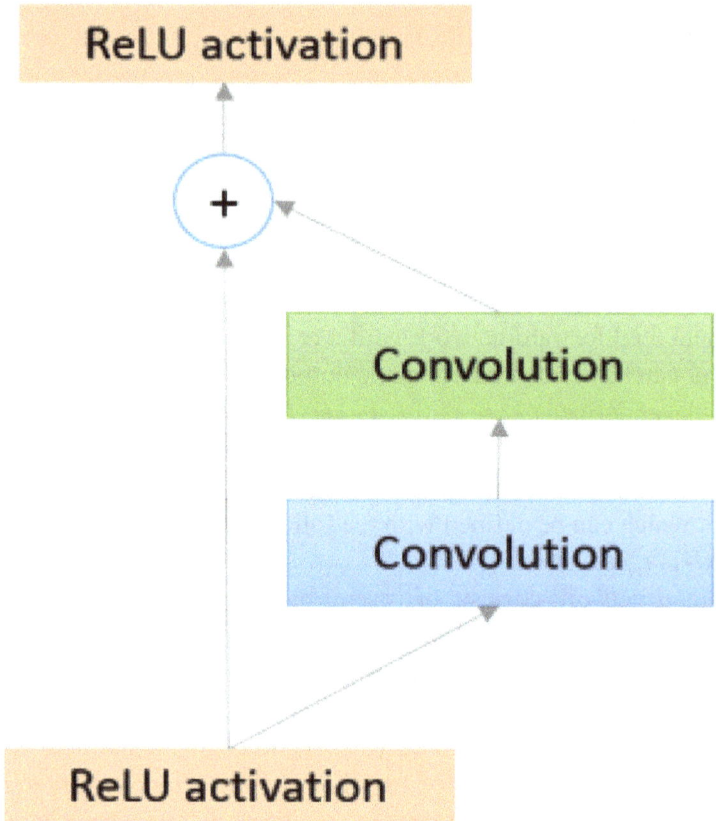

GoogLeNet (2014)

GoogLeNet, the winner of ILSVRC 2014 (Alom, 2018), was a model proposed by Christian Szegedy of Google with the objective of reducing computation complexity compared to the traditional CNN. The proposed method was to incorporate Inception Layers that had variable receptive fields, which were created by different kernel sizes. These receptive fields created operations that captured sparse correlation patterns in the new feature map stack.

The initial concept of the Inception layer can be seen in Figure 16. GoogLeNet improved state of-the-art recognition accuracy using a stack of Inception layers, seen in Figure 17. The difference between the naïve inception layer and final Inception Layer was the addition of 1x1 convolution kernels. These kernels allowed for dimensionality reduction before computationally expensive layers. GoogLeNet

Figure 16. The basic block diagram for inception Residual unit. (Alom, 2018)

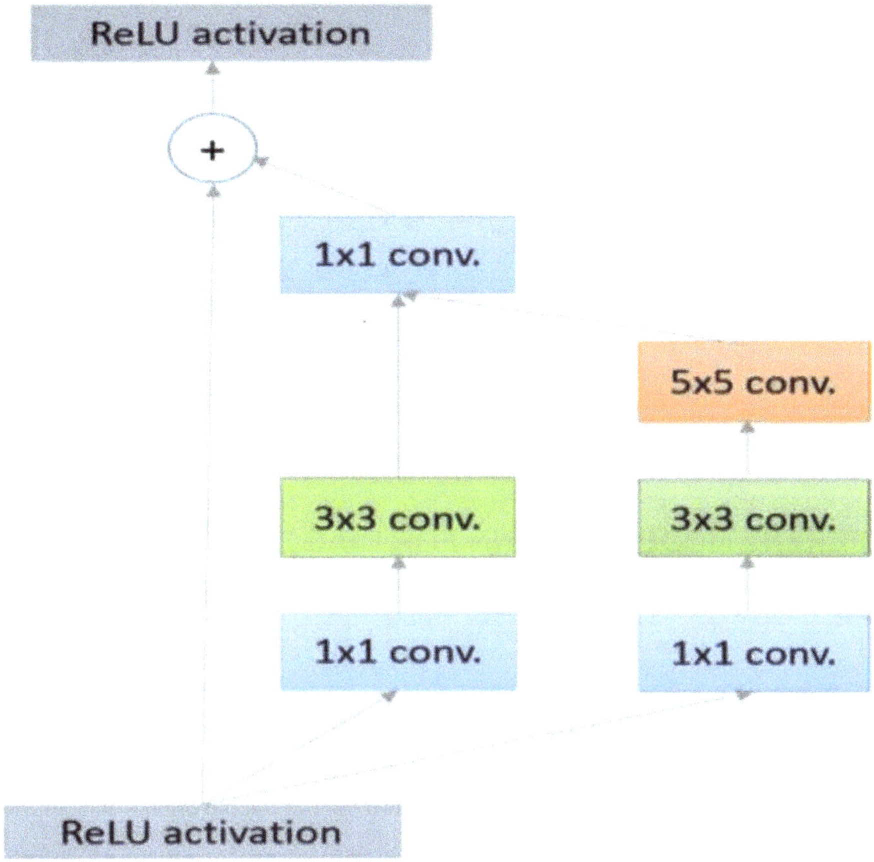

consisted of 22 layers in total, which was far greater than any network before it. Later improved version of this network is proposed in (Alom, 2018). However, the number of network parameters GoogLeNet used was much lower than its predecessor AlexNet or VGG. GoogLeNet had 7M network parameters when AlexNet had 60M and VGG-19 138M. The computations for GoogLeNet also were 1.53G MACs far lower than that of AlexNet or VGG.

Figure 17. Inception layer: Naïve version (Mansour, 2021)

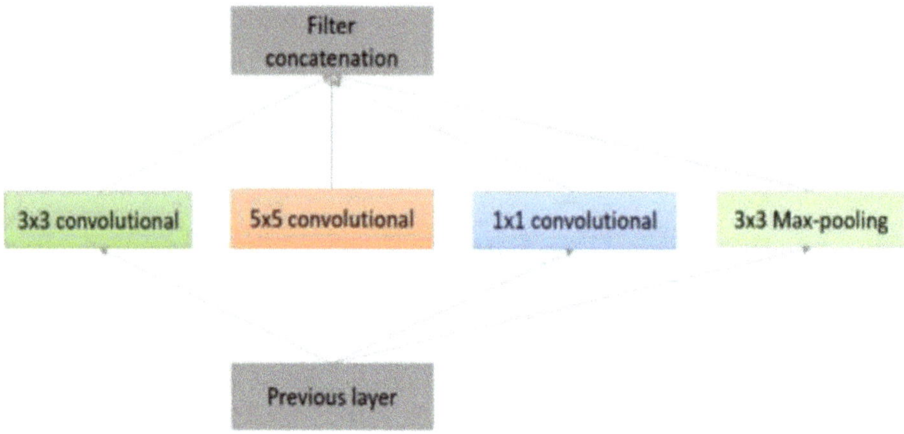

DIFFERENT MODELS OF CNN FOR IMAGES SEGMENTATION

U-Net

The U-Net has an encoder decoder structure, with jump connections between the corresponding encoding and decoding layers that allow the network to retain low-level functionality for final prediction. U-Net is a convolutional neural network developed for biomedical image segmentation at the Department of Computer Science, University of Freiburg, Germany.

Its architecture is based on the fully convolutional network, it has been modified and extended to work with fewer training images and to allow more accurate segmentation. Segmentation of a 512×512 image takes less than a second on a recent GPU. The network consists of a contracting part and an expanding path, which gives it a "U" shaped architecture. The contracting part is a typical convolution network which consists of repeated application of convolutions, each followed by a rectified linear unit (ReLU) and a maximum pooling operation. During contraction, spatial information is reduced while feature information is increased. The expansive path combines geographic and spatial feature information through a sequence of upward convolutions and concatenations with high-resolution features from the contracting path, so the U-Net architecture is separated into 3 parts:

1. The contraction / subsampling path
2. Bottleneck
3. The extension / oversampling channel

Figure 18. The U-net architecture (O'Shea, 2021)

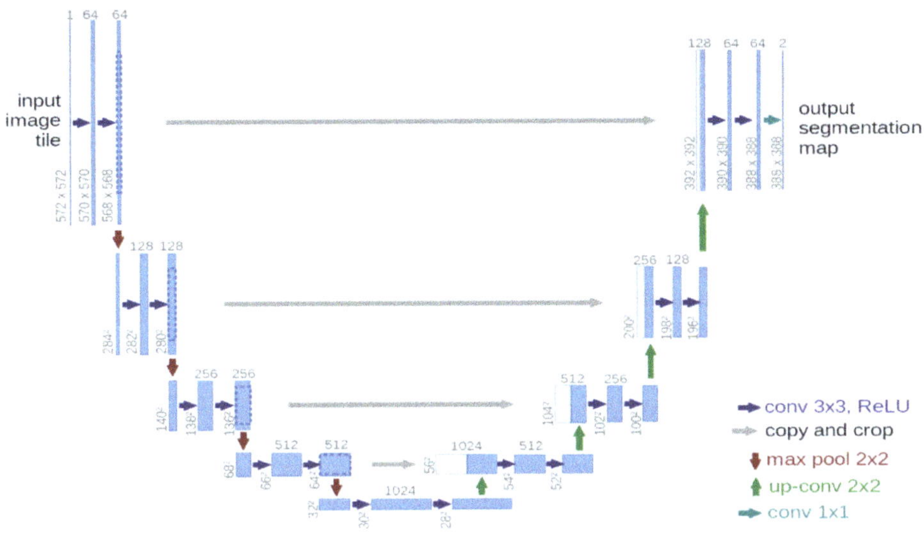

CONCLUSION

In recent years, object recognition is mainly based on the ML approach which gives high performance. A few years later, significant progress has been made in the field of object recognition, especially with the appearance of a new subfield called deep or deep learning. It is mainly based on the use of many neural networks of simple interconnected units in order to extract meaningful patterns from a large amount of data to solve complex problems such as: classification of medical images, detection of fraud, character recognition, etc.

In this chapter, we have detailed an explanation of the different basic concepts of deep learning, the classification models of the convolutional neural network and image segmentation.

REFERENCES

Abdurakhmonov, I. Y. (2016). *Bioinformatics: basics, development, and future.* InTech.

Al-Amri, S. S., & Kalyankar, N. V. (2010). Image segmentation by using threshold techniques. *Journal of Computing, 2*(5).

Alom, M. Z., Taha, T. M., Yakopcic, C., Westberg, S., Sidike, P., Nasrin, M. S., Hasan, M., Van Essen, B. C., Awwal, A. A. S., & Asari, V. K. (2019). A state-of-the-art survey on deep learning theory and architectures. *Electronics (Basel)*, *8*(3), 292. doi:10.3390/electronics8030292

Anderson, J. G., & Goodman, K. W. (Eds.). (2002). *Ethics and information technology: a case-based approach to a health care system in transition*. Springer New York.

Azar, A. T., & Eljamel, M. S. (2012). Medical robotics. In *Prototyping of robotic systems: applications of design and implementation* (pp. 253–287). IGI Global. doi:10.4018/978-1-4666-0176-5.ch009

Ball, J. E., Anderson, D. T., & Chan, C. S. Sr. (2017). Comprehensive survey of deep learning in remote sensing: Theories, tools, and challenges for the community. *Journal of Applied Remote Sensing*, *11*(4), 042609. doi:10.1117/1.JRS.11.042609

Basha, S. S., Dubey, S. R., Pulabaigari, V., & Mukherjee, S. (2020). Impact of fully connected layers on performance of convolutional neural networks for image classification. *Neurocomputing*, *378*, 112–119. doi:10.1016/j.neucom.2019.10.008

Chandrasekhar, K., Pavan, R., Bharathi, P., & Triveni, K. V. (2019). Histopathologic Cancer Detection. *IRJCS: International Research Journal of Computer Science*, *6*, 102–124.

Civantos, F. J., Stoeckli, S. J., Takes, R. P., Woolgar, J. A., De Bree, R., Paleri, V., Devaney, K. O., Rinaldo, A., Silver, C. E., Mondin, V., Werner, J. A., & Ferlito, A. (2010). What is the role of sentinel lymph node biopsy in the management of oral cancer in 2010? *European Archives of Oto-Rhino-Laryngology*, *267*(6), 839–844. doi:10.100700405-010-1215-1 PMID:20204392

Cumani, A. (1991). Edge detection in multispectral images. *CVGIP. Graphical Models and Image Processing*, *53*(1), 40–51. doi:10.1016/1049-9652(91)90018-F

Davis, J., & Goadrich, M. (2006, June). The relationship between Precision-Recall and ROC curves. In *Proceedings of the 23rd international conference on Machine learning* (pp. 233-240). 10.1145/1143844.1143874

Dehariya, V. K., Shrivastava, S. K., & Jain, R. C. (2010, November). Clustering of image data set using k-means and fuzzy k-means algorithms. In *2010 International conference on computational intelligence and communication networks* (pp. 386-391). IEEE. 10.1109/CICN.2010.80

Diehl, F. (2019). Edge contraction pooling for graph neural networks. *Cornell University ARIV.*

Duan, W., Zhang, L., Colman, J., Gulli, G., & Ye, X. (2021, September). Multi-modal Brain Segmentation Using Hyper-Fused Convolutional Neural Network. In *International Workshop on Machine Learning in Clinical Neuroimaging* (pp. 82-91). 10.1007/978-3-030-87586-2_9

Hong, J., Sapp, B., & Philbin, J. (2019). Rules of the road: Predicting driving behavior with a convolutional model of semantic interactions. In *Proceedings of the IEEE/CVF Conference on Computer Vision and Pattern Recognition* (pp. 8454-8462). 10.1109/CVPR.2019.00865

Jakhar, D., & Kaur, I. (2020). Artificial intelligence, machine learning and deep learning: Definitions and differences. *Clinical and Experimental Dermatology*, *45*(1), 131–132. doi:10.1111/ced.14029 PMID:31233628

Jing, J., Liu, S., Wang, G., Zhang, W., & Sun, C. (2022). Recent advances on image edge detection: A comprehensive review. *Neurocomputing*, *503*, 259–271. doi:10.1016/j.neucom.2022.06.083

Lee, J. H., Kim, D. H., Jeong, S. N., & Choi, S. H. (2018). Detection and diagnosis of dental caries using a deep learning-based convolutional neural network algorithm. *Journal of Dentistry*, *77*, 106–111. doi:10.1016/j.jdent.2018.07.015 PMID:30056118

Liu, Z., Mao, H., Wu, C. Y., Feichtenhofer, C., Darrell, T., & Xie, S. (2022). A convnet for the 2020s. In *Proceedings of the IEEE/CVF Conference on Computer Vision and Pattern Recognition* (pp. 11976-11986).

Mansour, R. F., & Aljehane, N. O. (2021). An optimal segmentation with deep learning based inception network model for intracranial hemorrhage diagnosis. *Neural Computing & Applications*, *33*(20), 13831–13843. doi:10.100700521-021-06020-8

Mateen, M., Wen, J., Song, S., & Huang, Z. (2018). Fundus image classification using VGG-19 architecture with PCA and SVD. *Symmetry*, *11*(1), 1. doi:10.3390ym11010001

Nwankpa, C., Ijomah, W., Gachagan, A., & Marshall, S. (2018). Activation functions: Comparison of trends in practice and research for deep learning. *Cornell University ARIV.*

O'Shea, R. J., Sharkey, A. R., Cook, G. J., & Goh, V. (2021). Systematic review of research design and reporting of imaging studies applying convolutional neural networks for radiological cancer diagnosis. *European Radiology*, *31*(10), 7969–7983. doi:10.100700330-021-07881-2 PMID:33860829

Rai, V., Quang, D. X., Erdos, M. R., Cusanovich, D. A., Daza, R. M., Narisu, N., Zou, L. S., Didion, J. P., Guan, Y., Shendure, J., Parker, S. C. J., & Collins, F. S. (2020). Single-cell ATAC-Seq in human pancreatic islets and deep learning upscaling of rare cells reveals cell-specific type 2 diabetes regulatory signatures. *Molecular Metabolism*, *32*, 109–121. doi:10.1016/j.molmet.2019.12.006 PMID:32029221

Sharma, N., Sharma, R., & Jindal, N. (2021). Machine learning and deep learning applications-a vision. *Global Transitions Proceedings*, *2*(1), 24–28. doi:10.1016/j.gltp.2021.01.004

Srivastava, S., Kumar, P., Chaudhry, V., & Singh, A. (2020). Detection of ovarian cyst in ultrasound images using fine-tuned VGG-16 deep learning network. *SN Computer Science*, *1*(2), 1–8. doi:10.100742979-020-0109-6

Sultana, F., Sufian, A., & Dutta, P. (2018, November). Advancements in image classification using convolutional neural network. In *Fourth International Conference on Research in Computational Intelligence and Communication Networks (ICRCICN)* (pp. 122-129). IEEE. 10.1109/ICRCICN.2018.8718718

Chapter 6
A Hybrid Classification Algorithm and Its Application on Four Real-World Data Sets

Lamiaa M. El Bakrawy
Al-Azhar University, Egypt

Abeer S. Desuky
https://orcid.org/0000-0003-1661-9134
Al-Azhar University, Egypt

ABSTRACT

The aim of this chapter is to propose a hybrid classification algorithm based on particle swarm optimization (PSO) to enhance the generalization performance of the adaptive boosting (AdaBoost) algorithm. AdaBoost enhances any given machine learning algorithm performance by producing some weak classifiers which requires more time and memory and may not give the best classification accuracy. For this purpose, PSO is proposed as a post optimization procedure for the resulted weak classifiers and removes the redundant classifiers. The experiments were conducted on the basis of ionosphere data set, thoracic surgery data set, blood transfusion service center data set (btsc) and Statlog (Australian credit approval) data set. The experimental results show that a given boosted classifier with post optimization based on PSO improves the classification accuracy for all used data. Also, the experiments show that the proposed algorithm outperforms other techniques with best generalization.

DOI: 10.4018/978-1-6684-5656-9.ch006

INTRODUCTION

Nowadays there is tremendous amount of data being collected and stored in databases everywhere across our realm. It is easy now to find databases with Terabytes - about 1,099,511,627,776 bytes - of data in enterprises and research fields. Numerous invaluable information and knowledge is buried in such databases; and without facile methods for extracting this buried information it is practically impossible to mine for them. Many algorithms were created throughout the decades for extracting what is called nuggets of knowledge from large sets of data. There are several diverse methodologies to approach this problem: classification, clustering, association rule, etc. our chapter will focus on classification (Sanakal & Jayakumari, 2014; Witten & Frank, 2005).

Classification is one of the most frequently studied problems by data mining and machine learning researchers (Sanakal & Jayakumari, 2014). Classification consists of predicting a certain outcome based on a given input. A classifier is a function or an algorithm that maps every possible input (from a legal set of inputs) to a finite set of classes or categories (Dhande & Dandekar, 2011). Adaptive boosting (AdaBoost) (Freund & Schapire, 1996) is a widespread successful technique used to boost the classification performance of weak learner. (Hu et al., 2014) proposed two algorithms based on AdaBoost classifier for online intrusion detection. They used the traditional AdaBoost where decision stumps are used as weak classifiers in the first algorithm. In the second algorithm, online Gaussian mixture models (GMMs) are used as weak classifiers to improve online AdaBoost process. The second algorithm showed a better performance in the experiments than the traditional AdaBoost process that uses decision stumps. Another improved AdaBoost algorithm named (ISABoost) proposed by (Qian et al., 2013) and applied in scene categorization. In ISABoost the inner structure of each trained weak classifier is adjusted before the traditional weights determination process. ISABoost algorithm after inner structure adjusting in each iteration of AdaBoost learning selects an optimal weak classifier and determines its weight. Three scene data sets used in Comparisons of ISABoost and traditional AdaBoost algorithms, where Back-propagation net- works and SVM are served as weak classifiers, and ISABoost verified its effectiveness.

(Choi et al., 2012) presented a novel multiple classifier system termed: “classifier ensemble”- based on AdaBoost for tackling false-positive (FP) reduction problem in Computer- aided Detection (CADe) systems, especially of mass abnormalities on Mammograms. Different feature representations were combined with data resampling based on AdaBoost learning to create the “classifier ensemble”. Adjusting the size of a resampled set is the effective mechanism used by classifier ensemble to regulate the degree of weakness of the weak classifiers of conventional AdaBoost ensemble. Support vector machines (SVM) and neural network (NN) with back-propagation algorithm were used as base classifiers and applied on digital database for screening mammography (DDSM) DB. The area under the receiver operating characteristics (ROC) was the used criterion to evaluate the classification performance and the comparative results showed the potential clinical effectiveness of the proposed ensemble. As the AdaBoost approach produces a large number of weak classifiers, particle swarm optimization (PSO) (Kennedy & Eberhart, 1995) has the potential to automatically elect a good set of weak classifiers for AdaBoost and improve the algorithm performance. Our goal is to optimize the AdaBoost algorithm performance using Particle Swarm Optimization technique.

BACKGROUND

Optimization techniques have been used for various real-world applications, (Hu et al., 2019) integrated PSO technique with the grey wolf optimizer (GWO) (Mirjalili et al., 2014) to classify endometrial carcinoma with gene expression by optimizing the parameters of the Elman neural network. In (Ganguly, 2020), the author presented a framework for simultaneous optimization of Distributed Generation (DG) penetration level and network performance index to determine the optimal sites, sizes and numbers of DG units. He utilized Multi-objective particle swarm optimization technique and validated it on a distribution system comprising 38 nodes. (Negi et al., 2021) implemented a hybrid approach with Particle Swarm Optimization (PSO) and Gray Wolf Optimization (GWO) to solve optimization problems named reliability allocation of life support systems and complex bridge systems. (Das et al., 2021), developed a modified genetic algorithm (MGA) by using a new selection technique, called a generation-dependent mutation and an in-vitro-fertilization-based crossover. The proposed technique was applied in commercial ship scheduling, supply environments and routing with dynamic demand. Their model could reduce risks and minimize port time with a static load factor. (Carvajal et al. 2021) applied a general type-2 fuzzy classification method for the medical assistance and the optimization of the general type-2 membership functions parameters using Ant Lion Optimizer (ALO) (Mirjalili, 2015) for comparison of type 1 and interval type-2 fuzzy classifiers with the Framingham dataset. The authors applied a general type-2 fuzzy classifier on the Jetson Nano hardware Development Board and compared the execution time of the type-1 and type-2 fuzzy classification technique. (Precup et al., 2021) proposed a new metaheuristic method called the Slime Mold Algorithm (SMA). The authors offered a fuzzy controller tuning technique and applied the concept of improving the performance of metaheuristics with information feedback techniques. The cost-effective fuzzy controllers and their tuning methods were validated in real-time with angular position control of the laboratory servo framework. (Valdez et al., 2020) presented a survey on Type-2 fuzzy logic controllers devised using nature-inspired optimization techniques. The authors exploited the most well-known optimizers to get the significant parameters on Type-2 and Type-1 fuzzy controllers to improve the gained outcome.

Data Mining Techniques

Several techniques such as Classification, Clustering, Regression, Artificial Intelligence, Neural Networks, Association Rules, Decision Trees, Genetic Algorithms, Nearest Neighbor method, and others are employed for knowledge discovery from databases (Ramageri, 2010).

Classification

Classification is one of the data mining technique, which utilizes a group of pre-classified samples to produce a model that can classify the entire population of information (Ramageri, 2010). This technique is particularly well suited to disease detection and prediction and frequently employs classification algorithms based on neural networks or decision trees. The process of data classification includes learning and classification. The training data are examined by a classification algorithm in Learning. Classification test results are used to estimate the classification rules' accuracy. The rules can be applied to new data sets if the accuracy is acceptable. This would contain whole records of both healthy people and patients evaluated on a record-by-record basis. These pre-classified samples are utilized by the classifier-training technique to find the set of parameters required for proper discrimination. Following that, these parameters are encoded by the technique into a model known as a classifier. The classification models contain:

a) Classification Based on Associations
b) Classification by decision tree induction
c) Neural Networks
d) Support Vector Machines (SVM)
e) Bayesian Classification

Clustering

Clustering is the method of identifying objects that belong to similar groups. Therefore, clustering algorithms can discover distribution patterns and relationships among large data sets, as well as locate dense and sparse regions in object space. Classification can be used to distinguish groups or classes of objects, but it is time-consuming, therefore clustering can be used as a preprocessing method for attribute subset selection and classification. For example, to categorize genes with similar functioning and generate groups of customers based on purchase patterns (Ramageri, 2010).

Types of clustering methods are:

a) Model-based methods
b) Density-based methods
c) Partitioning Methods
d) Hierarchical Agglomerative (divisive) methods
e) Grid-based methods

Prediction

Prediction can be done using regression techniques. One or more independent variables and dependent variables can be modeled using regression analysis. In data mining, independent variables are traits that are previously known, whereas responsive variables are the variables that we want to forecast. Regrettably, many real-world issues do not lend themselves to simple predictions. Sales volumes, stock prices, and product failure rates, for example, are all challenging to forecast due to the complicated interactions of various predictor factors. In order to estimate future values, more complicated techniques (such as logistic regression, decision trees, or neural networks) may be required. For both regression and classification, the same model types are commonly utilized. The CART (Classification and Regression Trees) decision tree algorithm, for example, can be used to create classification trees (to categorize categorical answer variables) as well as regression trees (to forecast continuous response variables). Both classification and regression models can be created using neural networks (Ramageri, 2010). Types of regression methods are:

a) Multivariate Nonlinear Regression
b) Linear Regression
c) Nonlinear Regression
d) Multivariate Linear Regression

Association Rule

The goal of association and correlation is to uncover common item set findings in huge data sets. This kind of information aids firms in making decisions like catalog design, cross-marketing, and customer shopping behavior analysis. Algorithms for association rules must be able to generate rules with confidence levels less than one (Ramageri, 2010). However, the number of feasible Association Rules for a particular dataset is usually quite vast, and a huge percentage of them are of little (if any) use. Types of association rules are:

a) Quantitative association rule
b) Multilevel association rule
c) Multidimensional association rule

Genetic Algorithm

Genetic algorithms are a sort of optimization method that is used to determine the maximum and minimum values of a function. Among the examples is a genetic algorithm that solves the famous Traveling Salesman problem. Genetic algorithms are a sort of optimization technique that is used to identify the best solution(s) to a given computing problem that maximizes or minimizes a particular metric. a specific function Genetic algorithms are a subset of the subject of study known as genetics. In that, they mirror the biological processes of reproduction, evolutionary computation (Kinnear, 1994), and using natural selection to find the 'best' solutions (Goldberg, 1989). The processes of a genetic algorithm are random, but this optimization technique allows one to control them. Set the randomization level and the control level (Goldberg, 1989). These algorithms are significantly more powerful and efficient than random search and exhaustive search techniques (Kinnear, 1994), and they don't require any additional problem knowledge. They can use this property to solve issues that conventional optimization methods can't solve because of a lack of continuity, derivatives, linearity, or other properties.

A lot of the language used in genetic algorithms comes from biology because they are meant to replicate biological processes. The entities that are referred to by this nomenclature in genetic algorithms, however, are significantly more straightforward than their biological equivalents (Mitchell, 1995). Almost all genetic algorithms share the following elements:

a) random mutation of chromosomes in the new generation
b) selection of which chromosomes will reproduce
c) a fitness function for optimization
d) crossover to produce the next generation of chromosomes
e) a population of chromosomes

The function that the algorithm is attempting to optimize is the fitness function. Evolutionary theory is the source of the word "fitness." The fitness function, which measures how "fit" each potential solution is, is the reason it is utilized in this situation. As one of the algorithm's most crucial components, the fitness function is covered in further detail at the end of this section. Chromosomes are a candidate answer to the issue that the genetic algorithm is attempting to solve, represented numerically by a value or values (Mitchell, 1995). Each candidate solution is encoded as an array of parameter values, a process that is also found in other optimization algorithms. If a problem has Npar dimensions, then typically each chromosome is encoded as an Npar-element array chromosome = [p1, p2, ..., pNpar]. where each pi is a particular value of the i-th parameter (Haupt & Haupt, 1998). The genetic algorithm's designer must figure out how to convert the sample space of potential solutions into chromosomes. The chromosomes can be made by, for example, converting each parameter value into a bit string (a series of 1s and 0s), then concatenating the bit strings end to end like genes in a DNA strand (Mitchell, 1995). This approach has historically been used to encode chromosomes, and it is still appropriate for discrete solution spaces. We will concentrate on binary chromosomes for the time being, while modern computers allow chromosomes to contain permutations, real numbers, and many other items.

The first generation of a genetic algorithm is made up of a randomly selected set of chromosomes (initial population). The fitness function then evaluates each chromosome in the population to see how well it addresses the challenge.

The selection operator now selects some of the chromosomes for reproduction depending on a probability distribution that the user defines. The more suitable a chromosome is, the more likely it will be chosen.

The selection operator selects chromosomes with replacement, which means that the same chromosome can be selected several times. In cell meiosis, the crossover operator resembles biological chromosome crossing and recombination. This operator creates two children by swapping a subsequence of two of the specified chromosomes. For instance, if the parent chromosomes are identical,

[11010111001000] and [01011101010010]

are crossed over after the fourth bit, then

[01010111001000] and [11011101010010]

will be their offspring.

The mutation operator randomly flips individual bits in the new chromosomes (turning a 0 into a 1 and vice versa). Typically, the mutation happens with a very low probability, such as 0.001. Some algorithms implement the mutation operator before the selection and crossover operators; this is a matter of preference. At first glance, the mutation operator may seem unnecessary. In fact, it plays an important role, even if it is secondary to those of selection and crossover. Selection and crossover maintain the genetic information of fitter chromosomes, but these chromosomes are only fitter relative to the current generation. This can cause the algorithm to converge too quickly and lose "potentially useful genetic material (1's or 0's at particular locations)" (Goldberg, 1989). In other words, the algorithm can get stuck at a local optimum before finding the global optimum (Haupt & Haupt, 2004). The mutation operator helps protect against this problem by maintaining diversity in the population, but it can also make the algorithm converge more slowly. Typically, the selection, crossover, and mutation process continue until the number of offspring is the same as the initial population so that the second generation is composed entirely of new offspring and the first generation is completely replaced. However, some algorithms let highly fit members of the first generation survive into the second generation. Now the second generation is tested by the fitness function, and the cycle repeats. It is a common practice to record the chromosome with the highest fitness (along with its fitness value) from each generation, or the "best-so-far" chromosome (Koza, 1994). Genetic algorithms are iterated until the fitness value of the "best-so-far" chromosome stabilizes and does not change for many generations. This means the algorithm has converted to a solution(s). The whole process of iterations is called a run. At the end of each run, there is usually at least one chromosome that is a highly fit solution to the original problem. Depending on how the algorithm is written, this could be the most fit of all the "best-so-far" chromosomes or the fittest of the final generation. The "performance" of a genetic algorithm depends highly on the method used to encode candidate solutions into chromosomes and "the particular criterion for success," or what the fitness function is actually measuring (Mitchell, 1995). Other important details are the probability of crossover, the probability of mutation, the size of the population, and the number of iterations. These values can be adjusted after assessing the algorithm's performance on a few trial runs. Genetic algorithms are used in a variety of applications. Some prominent examples are automatic programming and machine learning. They are also well suited to modeling phenomena in economics, ecology, the human immune system, population genetics, and social systems.

Decision Tree

A decision tree is a classification algorithm that is expressed as a recursive division of the instance space. The decision tree is made up of nodes that create a rooted tree, which is a directed tree with no incoming edges and a node named "root." Each of the other nodes has one incoming edge. An internal or test node has outgoing edges. All additional nodes are referred to as leaves (also known as terminal or decision nodes). Each internal node in a decision tree divides the instance space into two or more sub-spaces based on a discrete function of the input attribute values. Each test considers a single attribute in the simplest and most common situation, with the instance space partitioned according to the attribute's value. The condition refers to a range in the case of numeric properties. Each leaf is allocated to a class that represents the best target value. Alternatively, the leaf could contain a probability vector showing the likelihood that the target characteristic will have a specific value. Instances are categorized by traveling from the tree's root to a leaf's leaf, based on the results of the tests along the way. A decision tree is used to determine if a potential customer would respond to a direct mail piece. Internal nodes are shown as circles, while leaves are shown as triangular shapes. It's worth noting that this decision tree has both nominal and numeric attributes. The analyst can use this classifier to forecast a potential customer's response (by sorting it down the tree) and identify the behavioral characteristics of the complete potential customer population when it comes to direct mailing. Each node is labeled with the characteristic it is testing, and its branches are labeled with the values that correspond to that attribute. Decision trees can be mathematically understood as a collection of hyperplanes, each orthogonal to one of the axes, in the case of numeric characteristics. Naturally, decision-makers favor less complex decision trees since they are easier to understand. Furthermore, according to (Breiman et al., 1984), the intricacy of the tree has a significant impact on its correctness. The stopping criterion and pruning process are both expressly utilized to control the tree's complexity. The total number of nodes, total number of leaves, tree depth, and number of characteristics used are typically used to quantify tree complexity. Rule induction is closely connected to decision tree induction. Each path from a decision tree's root to one of its leaves can be converted into a rule by connecting the tests along the path to form the antecedent portion and using the leaf's class prediction as the class value.

PRELIMINARIES: ADABOOST AND PARTICLE SWARM OPTIMIZATION

AdaBoost

The AdaBoost algorithm is frequently used boosting method proposed by (Freund & Schapire, 1996). which was a kind of boosting algorithms proposed by (Schapire, 1990). AdaBoost algorithm is used to boost the classification performance of a weak learning classifier by combining iteratively a collection of weak classifiers to form a stronger classifier. The weak classifiers are combined by AdaBoost by taking into account a weight distribution on the training samples such that more weight is assigned to samples misclassified by the previous iterations (Cheng et al., 2010).

The AdaBoost algorithm takes as input a training set (x1, y1), (x2, y2), ..., (xm, ym) where each instance xi Î X, and each label y_i Î $\{-1, +1\}$ (Freund & Schapire, 1996; Cheng et al., 2010; Cao et al., 2014).

On each iteration $t = 1, \ldots, T$, a distribution Dt is computed over the m training instances where T is the number of weak classifiers c_t :

$$D_m^{(i)} = \frac{1}{m}, i=1,2, \ldots, m. \tag{1}$$

The given weak classifier or weak learning algorithm is applied in each iteration t to find a weak hypothesis

$$c_t : X \rightarrow \{-1, +1\} \tag{2}$$

Where the aim of the weak learner is to find a weak hypothesis with low weighted error ϵt relative to D_t

$$\epsilon_t = \sum_{i=1}^{m} D_t^{(i)}, [y_i \neq c_t(x_i)]. \tag{3}$$

then update the Distribution D_t over the m training examples in each iteration

$$D_{t+1}^{(i)} = \frac{D_t^{(i)} \exp(-\alpha_t\, y_i c_t(x_i))}{z_t}, i = 1,2,\ldots,m. \tag{4}$$

where,

$$\alpha_t = \frac{1}{2} \ln \frac{1-\epsilon_t}{\epsilon_t} \tag{5}$$

And z_t is a normalization factor. After T iterations the final strong classifier is:

$$c(X) = sign\left(\sum_{t=1}^{T} \alpha_t\, c_t(X)\right) \tag{6}$$

Particle Swarm Optimization

Particle swarm optimization (PSO) was proposed by (Kennedy & Eberhart, 1995). PSO is one of the latest evolutionary optimization techniques for continuous optimization problems. It simulates the social behavior of bird flocks or fish schools. In this social group, there is a leader who presents the best performance and guides the movement of the whole swarm. The movement of each particle is directed by the leader and its own knowledge. Thus, the behavior of each particle is a compromise between its individual memory and a collective memory (Wu, 2011; Qiu & Lau, 2014; Hasan et al., 2010).

The canonical PSO algorithm consists of a swarm of particles, which are initialized randomly with a population of candidate solutions. They move iteratively through the n-dimension problem space to search the new solutions, where the fitness function can be calculated as the certain qualities measure. Each particle i has a position represented by a position-vector $\overrightarrow{x_{ij}}$ (i and j are the index of the particle and its dimension), and a velocity represented by a velocity-vector $\overrightarrow{v_{ij}}$ (Soliman et al., 2012; Costa et al., 2010; Sun et al., 2014).In each iteration t, the update of the velocity from the previous velocity to the new velocity is calculated by Eq. (7).

$$v_{ij}\left(t+1\right)=\left\{wv_{ij}\left(t\right)+c_1r_1(P_{ij}\left(t\right)-x_{ij}\left(t\right)\right)+c_2c(G_j\left(t\right)-x_{ij}\left(t\right)). \quad (7)$$

where w is called the inertia weight, which governs how much the pervious velocity should be retained from the previous time step, r_1 and r_2 are random numbers chosen in the interval [0,1] for the j-th dimension of the i-th particle. c and c_2 is are positives constants called as social and cognitive coefficients, $P_{ij}\left(t\right)$ is the personal best (pbest) position, which is defined as a vector with the best fitness function value achieved by the particle in the past and $G_j\left(t\right)$ is the global best (gbest) is used to record the position with the best fitness function value achieved by the all the particles in the past. The new position is then determined by the sum of the previous position and the new velocity by Eq. (8).

$$x_{ij}\left(t+1\right)= x_{ij}\left(t\right))+v_{ij}\left(t+1\right). \quad (8)$$

From Eq. (7), a particle decides where to move next, considering its own experience, which is the memory of its best past position, and the experience of its most successful particle in the swarm. In the particle swarm algorithm, the particle

Figure 1. The proposed algorithm

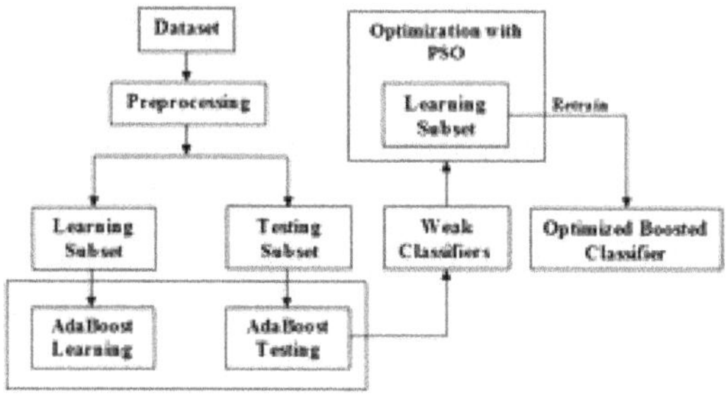

searches the solutions in the problem space with a range $\left[-s, s\right]$. In order to prevent the particle from flying away out of the search scope, the velocity is restricted on the interval $\left[-v_{max},\ v_{max}\right]$ given in Eq. (9):

$$v_{ij} = sign\left(v_{ij}\right)\min\left(\left|v_{ij}\right|, v_{max}\right). \tag{9}$$

Where the value of v_{max} is $p \times s$, with $0.1 \leq p \leq 1.0$ and is usually chosen to be s, i.e. p = 1. A given maximum number of iterations, number of iterations without improvement, or minimum fitness function error can be used as a stop criterion.

THE PROPOSED ALGORITHM

AdaBoost algorithm is a sequential forward search procedure based on the greedy selection strategy. Because of this strategy, the resulted weak classifiers and their coefficients are not optimal. We proposed PSO as a post optimization procedure for the resulted weak classifiers, removes the redundant classifiers and leads to shorter and better final classification performance.

According to the model of the AdaBoost algorithm that the weighted combination of weak learning classifiers $\left\{c_1,\ c_2,\ \ldots,\ c_r\right\}$ compose $c\left(X\right) = sign(\sum_{t=1}^{T} \propto_t c_t(X)$, the final boosted classifier. Unfortunately, the strong classifier comprises more weak classifiers requiring more memory and more time to evaluate and may produce less classification accuracy. To solve this, we propose a hybrid approach based on PSO as optimization algorithm. This algorithm (as shown in Figure 1.) elects the optimal weak classifiers with their weights as follows:

1st: Train the AdaBoost using the training data set. After T iterations of AdaBoost training, now we have T number of weak classifiers c_t with their weights $\propto_t$ which form the final strong classifier $\mathrm{c}(\mathrm{X})$, then the performance of the system with the testing data set is measured.

2nd: Initialize particle population, each particle in the evolving population is a binary vector q composed of $(q_1, q, \ldots, q_r)$ denoting the weak classifiers which constitute the final strong classifier. The particles move iteratively through the X-dimension, where each particle has a position set to 1 or 0 denoting the weak classifier appearance or absence which can be calculated according to the fitness function:

$$\mathrm{f}(\mathrm{q}) = 1 - E_q \tag{10}$$

where E_q is the error corresponding to the particle q. It is obvious that E_q expresses the fitness of q in the way that the smaller E_q is, the better q is. Each particle is updated in each iteration following the equations (7) (9).

3rd: In this step we use the resulted best binary vector Q to determine the final classifiers with their corresponding weights and then the optimal boosted classifier is calculated.

DATA SETS AND EXPERIMENTAL RESULTS

We have run our experiments using Matlab 12, on a system with a 2.30 GHZ Intel(R) Core(TM)i5 processor and 512 MB of RAM running Microsoft Windows 7 Professional (SP2).

The real-world data sets used throughout the chapter to our algorithm are: Ionosphere data set, Thoracic Surgery data set, Blood Transfusion Service Center data set (btsc) and Statlog (Australian Credit Approval) data set. Data sets are obtained from UCI Machine Learning Repository

(http://archive.ics.uci.edu/ml/datasets.html). A detailed description of the data sets is shown in Table I.

The parameters used for evolution were: In PSO, $c_1 = c_2 = 2.0$, $v_{max} = 4$, w was linearly decreased from 0.9 to 0.4, The PSO terminates if there is no found better individual within the next 10 generations.

From the experiment results shown in Table II, we can see that the accuracy of the classification for Ionosphere data set is increased by 1.24% due to the Particle Swarm Optimization. The accuracy of the classification for Thoracic Surgery data

Table 1. Properties of data sets

Properties of data sets	Ionosphere	Thoracic Surgery	btsc	Statlog
No. Of classes	2	2	2	2
No. Of examples	351	470	748	690
No. Of attributes	34	17	5	14

Table 2. Comparison of the boosted classifiers without and with post-optimization for ionosphere data set

AdaBoost classifiers	PSO AdaBoost accuracy	AdaBoost accuracy
8	87.8%	86.85%
16	87.92%	85.14%
24	88.03%	86.28%
32	88.51%	87.42%
40	88.76%	88.57%
48	89.25%	88.57%
Average	88.37%	87.13%

set and Blood Transfusion Service Center data set are also increased by 1.57% and 2.81% respectively, as shown in Tables III and IV. While Table V shows that accuracy of the classification for Statlog data set is slightly increased by 0.32%.

The results along with the comparison to other existing methods using Ionosphere data set, thoracic surgery data set, blood transfusion service center data set and Statlog data set are shown in Tables VI, VII, VIII and IX respectively. We found that generally the results achieved by the proposed method are higher than other techniques for all used data sets. The experimental results have demonstrated that a given boosted classifier with our post optimization based on particle swarm optimization increases the classification accuracy.

CONCLUSION

The main aim of this chapter is presenting a hybrid algorithm based on PSO to enhance the classification performance of the final classifier resulted from AdaBoost algorithm. Through the implementation of our algorithm, the experiments were conducted on the basis of four real-world data sets from the machine-learning

Table 3. Comparison of the boosted classifiers without and with post-optimization for thoracic surgery data set

AdaBoost classifiers	PSO AdaBoost accuracy	AdaBoost accuracy
8	83.6%	82.13%
16	81.87%	81.28%
24	80.68%	80.43%
32	82.33%	80%
40	82.11%	80.43%
48	81.77%	78.72%
Average	82.06%	80.49%

Table 4. Comparison of the boosted classifiers without and with post-optimization for blood transfusion service center data set

AdaBoost classifiers	PSO AdaBoost accuracy	AdaBoost accuracy
8	78.38%	78.34%
16	77.69%	71.39%
24	78.03%	73.52%
32	77.81%	74.86%
40	76.62%	74.86%
48	76.72%	75.40%
Average	77.54%	74.73%

Table 5. Comparison of the boosted classifiers without and with post-optimization for statlog data set

AdaBoost classifiers	PSO AdaBoost accuracy	AdaBoost accuracy
8	83.94%	83.76%
16	83.82%	83.47%
24	84.46%	83.76%
32	83.76%	84.92%
40	85.97%	84.92%
48	86.02%	85.21%
Average	84.66%	84.34%

Table 6. Classification accuracy comparison of machine learning algorithms using ionosphere data set

classifiers	Classification Accuracy
BP Neural Network Classifier (Dhande et al., 2012)	81.75%
RBF Neural Network (Dhande et al., 2012	82.78%
SVM Classifier (Dhande et al., 2012	84.46%
PSO SVM (Dhande et al., 2012	82.78%
K-nn (Hacibeyoglu et al., 2011)	86.3%
Naive Bayes (Hacibeyoglu et al., 2011)	87.8%
PSO AdaBoost(proposed)	88.37%

repository of University of California. The experimental results showed that a given boosted classifier with our post optimization based on particle swarm optimization performed quite well and improved the classification accuracy for all four data sets used with maximum accuracy increasing 2.81% for blood transfusion service center data and minimum accuracy increasing 0.32% for Statlog data set. The experiments also showed that the proposed algorithm outperforms other techniques applied on the same data. In future work, the proposed algorithm can be used to solve the imbalance problem in the data and/or improving other classifiers parameters and compare all improved classifiers to select the best performance. Moreover, other recent optimization algorithms can be utilized in comparison with PSO algorithm.

Table 7. Classification accuracy comparison of machine learning algorithms using thoracic surgery data set

classifiers	Classification Accuracy
PART (Sindhu et al., 2014)	76.59%
Na¨ıve Bayes (Harun & Alam, 2015)	77.74%
Multilayer Perceptron Harun & Alam, 2015)	80.91%
Boosted Na¨ıve Bayes Harun & Alam, 2015)	78.32%
Boosted Multilayer Perceptron Harun & Alam, 2015)	80.70%
Boosted J48 Harun & Alam, 2015)	79.34%
PSO AdaBoost(proposed)	82.06%

Table 8. Classification accuracy comparison of machine learning algorithms using blood transfusion service center data set

classifiers	Classification Accuracy
Online discretization (Vora & Mehta, 2012)	75.63%
CAIM (Vora & Mehta, 2012)	75.63%
Modified CAIM without merging stage (Vora & Mehta, 2012)	75.63%
Modified CAIM with`merging (Vora & Mehta, 2012)	75.63%
LR (Wang et al., 2015)	77.14%
NN (Wang et al., 2015)	75.55%
ELM (Wang et al., 2015)	76.20%
GBM (Wang et al., 2015)	76.34%
RF (Wang et al., 2015)	75.05%
PSO AdaBoost(proposed)	77.54%

Table 9. Classification accuracy comparison of machine learning algorithms using statlog data set

classifiers	Classification Accuracy
IDE3 (Anyanwu & Shiva, 2019)	71.5%
C4.5 (Anyanwu & Shiva, 2019)	84.2%
AdaboostC4.5 (Ferdousy et al., 2013)	84.01%
Bagging NB (Ferdousy et al., 2013)	77.81%
Adaboost NB (Ferdousy et al., 2013)	81.16%
MAdaBoost SVM (Wang et al., 2009)	81.20%
Single SVM (Wang et al., 2009)	79.86%
Arc-x4 SVM (Wang et al., 2009)	78.94%
AdaBoost SVM (Wang et al., 2009)	79.16%
K-nn (Hacibeyoglu et al., 2011)	57.5%
CN2 (Hacibeyoglu et al., 2011)	84.2%
PSO AdaBoost (proposed)	84.66%

REFERENCES

Anyanwu, M. & Shiva, S. (2009). Comparative Analysis of Serial Decision Tree

Classification Algorithms. *International Journal of Computer Science and Security, (IJCSS)*, *3*(3).

Breiman, L., Friedman, J., Olshen, R., & Stone, C. (1984). *Classification and Regression Trees*. Chapman and Hall.

Cao, J., Chen, J., & Li, H. (2014). An Adaboost- Backpropagation Neural Network for Automated Image Sentiment Classification. *TheScientificWorldJournal, 2014*, 1–9. doi:10.1155/2014/364649 PMID:25162047

Carvajal, O., Melin, P., Miramontes, I., & Prado-Arechiga, G. (2021). Optimal design of a general type-2 fuzzy classifier for the pulse level and its hardware implementation. *Engineering Applications of Artificial Intelligence*, *97*, 104069. doi:10.1016/j.engappai.2020.104069

Cheng, Z., Zhang, Y., Zhou, C., Zhang, W. & Gao, S. (2010). Classification of Skin Sensitizers on the Basis of Their Effective Concentration 3 Values by Using Adaptive Boosting Method. *International Journal of Digital Content Technology and its Applications (JDCTA)*, *4*(2), 109 – 121.

Choi, J. Y., Kim, D. H., Plataniotis, K. N., & Ro, Y. M. (2012). *Combining Multiple Feature Representations and AdaBoost Ensemble Learning for Reducing False-Positive Detections in Computer-aided Detection of Masses on Mammograms.* 34th Annual International Conference of the IEEE EMBS. San Diego, California USA.

Costa, J. S., Nadia, N. N., & Mourelle, L. (2010). Automatic Adaptive Modeling of Fuzzy Systems Using Particle Swarm Optimization. *Transactions on Computational Science VIII. Lecture Notes in Computer Science*, *6260*, 71–84. doi:10.1007/978-3-642-16236-7_5

Das, M., Roy, A., Maity, S., Kar, S., & Sengupta, S. (2021). Solving fuzzy dynamic ship routing and scheduling problem through new genetic algorithm. *Decision Making: Applications in Management and Engineering*. doi:. doi:10.31181/dmame181221030d

Dhande, J. D., & Dandekar, D. R. (2011). PSO Based SVM as an Optimal Classifier for Classification of Radar Returns from Ionosphere. *International Journal on Emerging Technologies*, *2*(2), 1–3.

Dhande, M., Dandekar, D., & Badjate, S. (2012). Performance Improvement of Ann Classifiers using Pso. *Proceedings of National Conference on Innovative Paradigms in Engineering & Technology,* (pp. 32-36). International Journal of Computer Applications.

Ferdousy, E., Islam, M. & Matin, A. (2013). Combination of Na¨ıve Bayes Classifier and K-Nearest Neighbor (cNK) in the Classification Based Predictive Models. *Computer and Information Science, 6*(3).

Freund, Y., & Schapire, R. E. (1996). Experiments with a new boosting algorithm. In *Proceedings of the 13th International Conference on Machine Learning,* (pp. 148-156).

Ganguly, S. (2020). Multi-objective distributed generation penetration planning with load model using particle swarm optimization. *Decis. Mak. Appl. Manag. Eng., 3*(1), 30–42. doi:10.31181/dmame2003065g

Goldberg, D. E. (1989). *Genetic Algorithms in Search, Optimization, and Machine Learning*. Addison-Wesley.

Hacibeyoglu, M., Arslan, A. & Kahraman, S. (2011). Improving Classification Accuracy with Discretization on Datasets Including Continuous Valued Features. *International Scholarly and Scientific Research & Innovation, 5*(6).

Harun, A., & Alam, N. (2015). Predicting Outcome of Thoracic Surgery by Data Mining Techniques. *International Journal of Advanced Research in Computer Science and Software Engineering, 5*(1).

Hasan, S., Shamsuddin, S., & Yusob, B. (2010). Enhanced Self Organizing Map (SOM) and Particle Swarm Optimization (PSO) for Classification. *Jurnal Generic, 5*, 7–11.

Haupt, R. L., & Haupt, S. E. (1998). *Practical Genetic Algorithms*. Wiley Interscience.

Haupt, R. L., & Haupt, S. E. (2004). *Practical Genetic Algorithms* (2nd ed.). Wiley.

Hu, H., Wang, H., Bai, Y., & Liu, M. (2019). Determination of endometrial carcinoma with gene expression based on optimized Elman neural network. *Applied Mathematics and Computation, 341*, 204–214. doi:10.1016/j.amc.2018.09.005

Hu, W., Gao, J., Wang, Y., Wu, O., & Maybank S.J. (2014). Online Adaboost Based Parameterized Methods for Dynamic Distributed Network Intrusion Detection. *IEEE T. Cybernetics*, (pp. 66-82).

Kennedy, J., & Eberhart, R. ((1995). Particle swarm optimization. *In Proceedings International Conference on Neural Networks,* (pp. 1942-1948).

Kinnear, K. E. (1994). A Perspective on the Work in this Book. K. E. Kinnear (Ed.), Advances in Genetic Programming (pp. 3-17). Cambridge: MIT Press.

Koza, J. R. (1994). Introduction to Genetic Programming. In K. E. Kinnear (Ed.), *Advances in Genetic Programming* (pp. 21–41). MIT Press.

Mirjalili, S. (2015). The Ant Lion Optimizer. *Advances in Engineering Software*, *83*, 80–98. doi:10.1016/j.advengsoft.2015.01.010

Mirjalili, S., Mirjalili, S. M., & Lewis, A. (2014). Grey Wolf Optimizer. *Advances in Engineering Software*, *69*, 46–61. doi:10.1016/j.advengsoft.2013.12.007

Mitchell, M. (1995). Genetic Algorithms: An Overview. *Complexity*, *1*(1), 31–39. doi:10.1002/cplx.6130010108

Negi, G., Kumar, A., Pant, S., & Ram, M. (2021). Optimization of Complex System Reliability using Hybrid Grey Wolf Optimizer. *Decis. Mak. Appl. Manag. Eng.*, *4*(2), 241–256. doi:10.31181/dmame210402241n

Precup, R.-E., David, R.-C., Roman, R.-C., Petriu, E. M., & Szedlak-Stinean, A.-I. (2021). Slime Mould Algorithm-Based Tuning of Cost-Effective Fuzzy Controllers for Servo Systems. *Int. J. Comput. Intell. Syst.*, *14*(1), 1042–1052. doi:10.2991/ijcis.d.210309.001

Qian, X., Tang, Y. Y., Yan, Z., & Hang, K. (2013). ISABoost: A weak classifier inner structure adjusting based AdaBoost algorithm-ISABoost based application in scene categorization. *Neurocomputing 103,* (pp. 104-113). Elsevier.

Qiu, X., & Lau, H. (2014). An AIS-based hybrid algorithm for static job shop scheduling problema. *Journal of Intelligent Manufacturing*, *25*(3), 489–503. doi:10.100710845-012-0701-2

Ramageri, B. M. (2010). Data Mining Techniques and Applications. *Indian Journal of Computer Science and Engineering*, *1*(4), 301–305.

Sanakal, R., & Jayakumari, S. T. (2014). Prognosis of Diabetes Using Data mining Approach-Fuzzy C Means Clustering and Support Vector Machine. *International Journal of Computer Trends and Technology*, *11*(2), 94–98. doi:10.14445/22312803/IJCTT-V11P120

Schapire, R. E. (1990). The Strength of Weak Learnability. *Machine Learning*, *5*(2), 197–227. doi:10.1007/BF00116037

Sindhu, V., Prabha, S., Veni, S., & Hemalatha, M. (2014). Thoracic Surgery Analysis Using Data Mining Techniques. *Int.J. Computer Technology and Application*, *5*(2), 578–586.

Soliman, M., Hassanien, A., Ghali, N., & Onsi, H. (2012). An adaptive Watermarking Approach for Medical Imaging Using Swarm Intelligent. *International Journal of Smart Home*, *6*(1), 37–50.

Sun, J., Palade, V., Cai, Y., Fang, W., & Wu, X. (2014). Biochemical systems identification by a random drift particle swarm optimization approach. *BMC Bioinformatics 15*(Suppl 6), 1-17. http://www.biomedcentral.com/1471- 2105/15/S6/S1.

Valdez, F., Castillo, O., Cortes-Antonio, P., & Melin, P. (2020). A survey of Type-2 fuzzy logic controller design using nature inspired optimization. *Journal of Intelligent & Fuzzy Systems, 39*(5), 6169–6179. doi:10.3233/JIFS-189087

Vora, S., & Mehta, R. (2012). MCAIM: Modified CAIM Discretization Algorithm for Classification. *International Journal of Applied Information Systems,* 3(5).

Wang, S., Mathew, A., Chen, Y., Xi, L., Ma, L., & Lee, J. (2009). Empirical analysis of support vector machine ensemble classifiers. *Expert Systems with Applications*, *36*(3), 6466–6476. doi:10.1016/j.eswa.2008.07.041

Wang, Y., Li, Y., Xiong, M., & Jin, L. (2015). Random Bits Regression: a Strong General Predictor for Big Data.

Witten, I. H., & Frank, E. (2005). Data Mining: Practical Machine Learning Tools and Techniques (2nd ed.). Morgan Kaufmann Publishers, Elsevier Inc.

Wu, J. (2011). Integrated Real-Coded Genetic Algorithm and Particle Swarm Optimization for Solving Constrained Global Optimization Problems. *Advances in Information Technology and Education Communications in Computer and Information Science*, *201*, 511–522. doi:10.1007/978-3-642-22418-8_72

KEY TERMS AND DEFINITIONS

Accuracy: Refers to the degree of closeness between a measurement and its accepted or true value.

Data Mining: The computer-assisted process of analyzing dense volumes of data and extracting interesting and useful information.

Dataset: A set (collection) of related (discrete) items that may be managed independently or in groups or accessed as a whole entity.

Fitness Function: A function that takes an elected solution to the problem as an input and yields as an output how the solution is fit.

Machine Learning: The scientific study of using computer systems and improving it to be able to learn and adapt without using an explicit instruction.

Optimization Algorithms: Search techniques that are used to find an optimal solution to an optimization problem, possibly subject to a set of constraints.

UCI Machine Learning Repository: A set (collection) of domain theories, databases and data generators which are utilized for experimental analysis of machine learning algorithms by machine learning researchers.

Chapter 7
Advanced Machine Learning Techniques in Spam Medical Events:
Review of Literature

Yasmin Bouarara
Dr. Moulay Tahar University, Algeria

ABSTRACT

The detection of adverse events (spam) within a health establishment, associated or not with care, represents one of the axes of a risk management approach. It does not in any way constitute a means of denunciation, control, or sanction; the purpose being that the most important medical spam events do not recur. In this chapter the authors discuss the different algorithms of machine learning such as the KNN decision tree, naive bayes, etc., applicable to the filtering of medical spam event. The objective of these techniques is to control medical data in order to make decisions and achieve strategic objectives.

INTRODUCTION

The term spam health event includes all abnormal emails situations that impact people's health: infections, accidents related to care, health threats, etc. Their signaling (declaration) promotes the identification of risks and the implementation of measures aimed at correcting the problem and, whenever possible, preventing its recurrence.

After the mailbox of our home, it is now our electronic mailbox with the explosion of Internet use. Nowadays, social networking sites such as Facebook, Myspace, and

DOI: 10.4018/978-1-6684-5656-9.ch007

Twitter have become one of the main vectors for users to keep track and communicate with their friends online. Merely, the number of electronic mail box is increasing. Each user has at least an email address; the minimum number of BALE (Box for Electronic Arts) is 800 million worldwide. Such mass approachable person is of course a boon to advertisers, but also a favored means of communication for the spammers, scammers, hackers, political, publicity.... etc.

In this chapter we will see an overview concerning the preliminary concepts of spam event and the different detection techniques existed in literature.

Spam History

The real origin of the term "SPAM" comes from 1970 Monty Python's Flying Circus skit. In this skit, all the restaurant's menu items devolve into SPAM. When the waitress repeats the word SPAM, a group of Vikings in the corner sing "SPAM, SPAM, lovely SPAM, Wonderful SPAM" drowning out other conversation, until they are finally told to shut it.

Although the first spam message had already been sent via telegram in 1864, then it was sent as commercial e-mail occurred in 1978, the term spam for this practice had not yet been applied in the 1980s. It was adopted to describe certain users who frequented BB (Bulletin board is a computer system running software that allows users to dial into the system over a phone line or Telnet), who would repeat "SPAM" a huge number of times to scroll other users' text off the screen in early chat rooms services like the early days of AOL (Delany, 2012).

Spam

Spam is considered to be an unsolicited commercial electronic message (figure 2). It is often a source of scams, computer viruses, and offensive content that takes up valuable time and increases costs for consumers, businesses, and governments (Dada, 2019).

The Different Types of Spam

The most common spam is of course linked to spam emails. Nevertheless, there are different forms of spam:

Figure 1. A model of spam email

Mail considère ce message comme indésirable

De : euroonlineresult <euroonlineresult@libero.it>
Objet : **[multimedia] Read This Mail Carefully**
Date : 14 novembre 2008 12:50:20 HNEC
À : undisclosed-recipients:;

Today, we announce the lucky winners of e-mail lottery held on the 12-11-08in spain youremail address won the sum of the sum of €1.000,000 your chip number9465206 Ref:M154S/WLO8/EU21 contact your asigned lawyer for further information,
Mr Teddy Williams. +34-658 654 308, Fax:+34912919713E-mail: teddywilliamsclaimsofficer@gmail.com,
Co-odinator,Sanchez
Laura

Spam Voice Over IP

The spam VoIP, also called SPIT or vishing. SPLIT is a new kind of spam via the telephone. It's like anonymous call issued at any time of day or night, issued to raise (as phishing technique) personal information (Saberi, 2007).

The Spam Messages in the Discussion Forums

This is an advertising message (containing commercial nature hyperlinks) left on some forums the goal is the same as the spam received by email: advertise for free (Saberi, 2007).

Spam in Blogs (SIG)

It is called SPLOG (contraction of spam and blog). It is a very popular technique it's to let Internet users on blogs with links to advertising sites (Fumera, 2007).

PHISHING

It is called *filoutage* or *hameçonnage* in French as presents the next figure 3. It is a technique by which attackers pose major corporations or financial institutions that are familiar by sending fraudulent e-mails. It retrieves passwords of bank accounts or credit card numbers. In this case the hacker could create a false social network

Figure 2. A phishing model (Chirita2005)

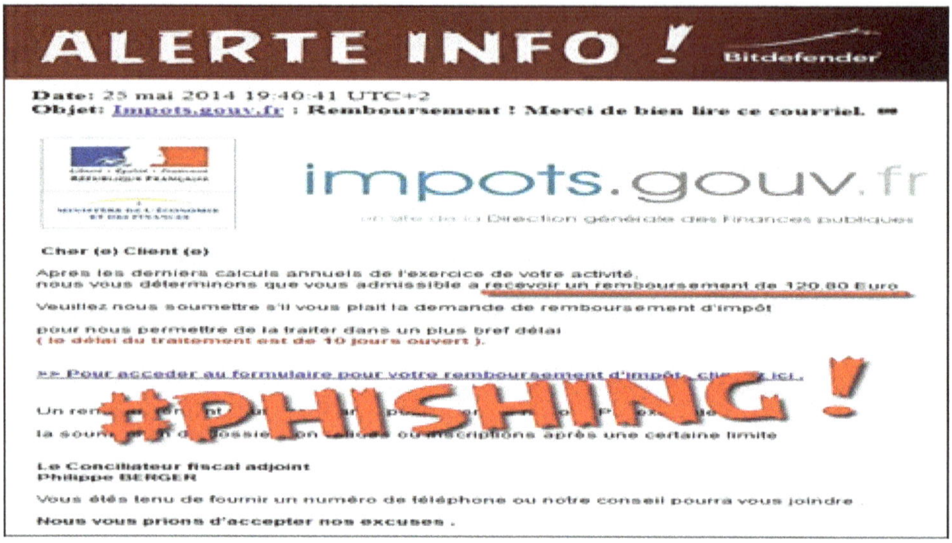

page (Facebook, Twitter, etc.) that appears entirely legitimate. Then, when you try to connect the fake page, it saves your information with your username and password in hand [A3].

Scam

It is based on using a false name or a false quality. It is a technique that abuses the credulity of email recipients to extort money. For example, someone proposes to a claim to have a large sum of money and wants to collaborate with him to use an existing bank account to transfer money. Of course, in exchange for this person service promises a percentage of the large sum to be channeled through profit account of the recipient (Grier, 2010).

SPAMMERS ACTIVITIES

There are many ways in which spammers can harvest your email address such as:

Image Spam

The spammers put their spam into the image, they put the text as a GIF or JPEG image like advertisement and business then they send it in email, and it's difficult to detect that kind of spam because they exist different type of images (color modified image, gray image, multi farm animated, image background, image text image...) (Fumera, 2007).

Zombie Spam

It's a collection of computer systems that send millions of spam messages at once. Nevertheless, it can find itself in bad reputation with the spams filters (Gaurav, 2020).

Java Script Tricky

It makes the messages more difficult to decode. The spammers' dispatch the contents of the spam message into java script in which will be activated once the message is opened (Firte, 2010).

Spam Creation Process

The different steps to create a spam can be defined as follows:

1. Collection and address verification: sort addresses by groups (targets)
2. Creating platforms for mass mailing (servers and / or individual computers)
3. Writing mass mailing programs
4. Promoting the spammer services
5. Development texts for specific campaigns then t it send the spam

Filtering

It's a program or a source of code that allowed to import or export data where it allow data to pass just if it's matched with the given information and when we speak about the spam filtering we are trying to find a solution to eliminate the spam or put it in the junk mail box it's also called a spam blockers it's configured and based on different and variety of criteria including the email senders and specific words whether in the subject or the message (Krasser, 2007).

I.8-Spam Detection Systems Existed in Literature

The most known spam detection techniques existed in literature are grouped in the next figure 4 and detailed in the next paragraphs:

Figure 3. Different approach of spam detection

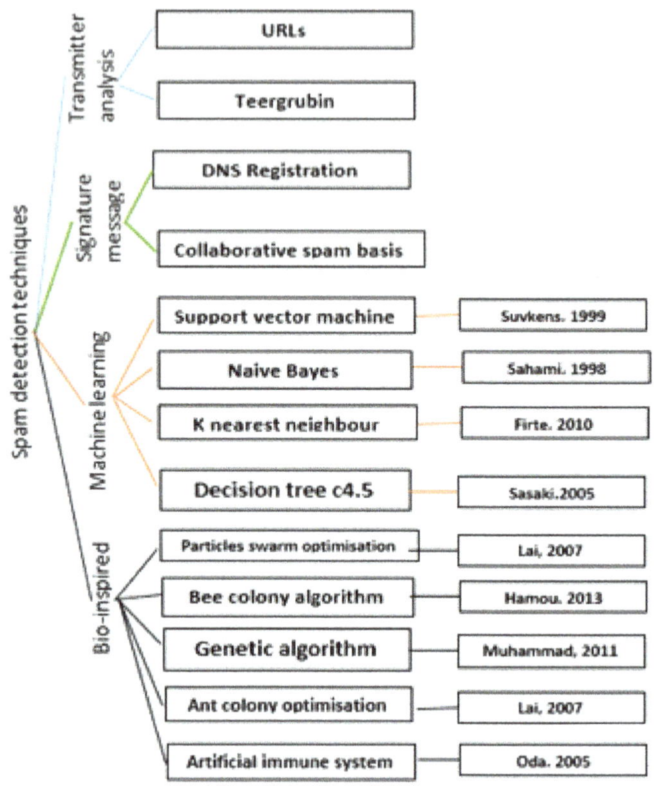

Classical Family for Spam Detection:

There are two large families of classical techniques to deal with spammers as detailed in the next titles:

Digital Signature family

This kind of techniques usually based on a set of digital signature. They need to have a detailed analysis of mails that circulate in the network. Among the methods according to this principle, we can distinguish DNS registration (Chirita, 2005), blacklists (Grier, 2010) and Grey list.

Transmitter Analysis

These techniques are based on the user profile and the analysis of the spammers' behaviour such as the methods URLs analysis in (Cormack, 2007).

Machine Learning Family for Spam Detection

Among the techniques found in the literature following the principle of this family as fellow:

Naive Bayes NB

The NB is a probabilistic model based on Bayes theorem and belonging to the linear algorithms group that has been used by Sahami in (Sahami, 1998) to improve the anti-spam solution.

K-Nearest Neighbours (KNN)

The KNN is a very simple algorithm that allows an instance-based classification using a similarity measure and fixing a parameter K. This algorithm was applied for the construction of an online application that can detect undesirable mail by Firte & al in (Firte, 2010).

Support Vectors Machine (SVM)

SVM is a classification technique that aims to find a function hyper plane based on the supports vector calculated by Lagrange to maximize the margin between classes. Ducker and his collaborators in (Suykens, 1999) used this principle for the categorization of electronic messages (spam and ham).

K-Means

K Means is an unsupervised algorithm based on the initialization of the parameter K. Sasaki and al in (Cormack, 2007) thought to apply this principle with K = 2 (spam and ham) for spam detection problem.

Decision Tree c4.5

After a new form of spamming based on image has appeared. In (Krasser, 2007) the researchers had applied the C4.5 to generate decision rules that will make a choice between the spam image and ham image.

Bio-Inspired Family for Spam Detection

This section is dedicated to the different work done around the spam detection using the bio-inspired techniques.

Genetic Algorithms for Spam Detection (GA-SD)

A very interesting paper was presented (Mohammad, 2011) as a recapitulation of the application of three bio-inspired algorithms (artificial immune system AIS, genetic algorithms and GAs artificial neural network ANN) for the detection of spam (Holland, 1973).

Particle Swarm Optimization for Spam Detection (PSO-SD)

Lai et al (Mohammad, 2011) had the idea of examining how particle swarm optimization algorithm can help select relevant features for discriminating the classification of solicited and unsolicited messages (Kennedy, 1995).

Ant Colony Optimization for Spam Detection (ACO-SD)

Taweessiwate et al. (2012) had the idea to reformulate the principle of ACO for the problem of host spam detection using the webspam-UK2006 dataset for his experimentation (Zhang, 2021).

Artificial Social Bees for Spam Detection (ASB-SD)

Zhang et al (2021) have formulated a new method inspired from the social life of bees, and their way to ensure the proper functioning of the hive using the filtering principle. Each filter is by ensuring bee workers (supervisor, cleaner and collector). A message is identified as spam if it succeeds spent three filtering (Rapacz, 2021).

Artificial Immune System for Spam Detection (AIS-SD)

To better counter the phenomenon of spamming the researchers considered spam as pathogens that are combated by humans where a spam detection method was introduced by Oda et al (Mohammad,2011) called artificial immune system AIS to effectively protect the email users. This algorithm has been tested on the Spam Assassin corpus and gave satisfactory results (Garg, 2021).

CONCLUSION

It is important for the administrators of electronic mail servers and organizer of information systems to have a better detector of spam. Newly, about 85% of information that exists in the web and the emails exchanged between the users, the business and administrators are in unstructured format (textual data). In order to better control the data and have a best filtering, we must have a better representation of emails to make them interpreted by the machine. This is the context of the next chapter which brings together the different steps of text mining and text pre-processing.

REFERENCES

Chirita, P. A., Nejdl, W., & Zamfir, C. (2005). Preventing shilling attacks in online recommender systems. In WIDM'05 (pp. 67–74). doi:10.1145/1097047.1097061

Dada, E. G., Bassi, J. S., Chiroma, H., Adetunmbi, A. O., & Ajibuwa, O. E. (2019). Machine learning for email spam filtering: Review, approaches and open research problems. *Heliyon*, *5*(6), e01802. doi:10.1016/j.heliyon.2019.e01802 PMID:31211254

Delany, S. J., Buckley, M., & Greene, D. (2012). SMS spam filtering: Methods and data. *Expert Systems with Applications*, *39*(10), 9899–9908. doi:10.1016/j.eswa.2012.02.053

Firte, L., Lemnaru, C., & Potolea, R. (2010, August). Spam detection filter using KNN algorithm and resampling. In *International Conference on Intelligent Computer Communication and Processing* (pp. 27-33). IEEE. 10.1109/ICCP.2010.5606466

Fumera, G., Pillai, I., Roli, F., & Biggio, B. (2007, March). Image spam filtering using textual and visual information. In *Proceedings of the MIT Spam Conference.*

Garg, P., & Girdhar, N. (2021, January). A Systematic Review on Spam Filtering Techniques based on Natural Language Processing Framework. In *11th International Conference on Cloud Computing, Data Science & Engineering (Confluence)* (pp. 30-35). IEEE. 10.1109/Confluence51648.2021.9377042

Gaurav, D., Tiwari, S. M., Goyal, A., Gandhi, N., & Abraham, A. (2020). Machine intelligence-based algorithms for spam filtering on document labeling. *Soft Computing*, *24*(13), 9625–9638. doi:10.100700500-019-04473-7

Grier, C., Thomas, K., Paxson, V., & Zhang, M. (2010, October). @ spam: the underground on 140 characters or less. In *Proceedings of the 17th ACM conference on Computer and communications security* (pp. 27-37). ACM 10.1145/1866307.1866311

Holland, J. H. (1973). Genetic algorithms and the optimal allocation of trials. *SIAM Journal on Computing*, *2*(2), 88–105. doi:10.1137/0202009

Kennedy, J., & Eberhart, R. (1995). Particle Swarm Optimization. *Proceedings of IEEE International Conference on Neural Networks*, pp. 1942–1948 10.1109/ICNN.1995.488968

Krasser, S., Tang, Y., Gould, J., Alperovitch, D., & Judge, P. (2007, June). Identifying image spam based on header and file properties using C4. 5 decision trees and support vector machine learning. In *Information Assurance and Security Workshop* (pp. 255-261). IEEE.

Mohammad, A. H., & Zitar, R. A. (2011). Application of genetic optimized artificial immune system and neural networks in spam detection. *Applied soft computing, 11*(4), 3827-3845.

Rapacz, S., Chołda, P., & Natkaniec, M. (2021). A Method for Fast Selection of Machine-Learning Classifiers for Spam Filtering. *Electronics (Basel)*, *10*(17), 2083. doi:10.3390/electronics10172083

Saberi, A., Vahidi, M., & Bidgoli, B. M. (2007, November). Learn to detect phishing scams using learning and ensemble? methods. In *Web Intelligence and Intelligent Agent Technology Workshops, 2007 IEEE/WIC/ACM International Conferences*, (pp. 311-314). IEEE.

Suykens, J. A., & Vandewalle, J. (1999). Least squares support vector machine classifiers. *Neural Processing Letters*, *9*(3), 293–300. doi:10.1023/A:1018628609742

Taweesiriwate, A., Manaskasemsak, B., & Rungsawang, A. (2012, March). Web spam detection using link-based ant colony optimization. In *International Conference on Advanced Information Networking and Applications (AINA)* (pp. 868-873). IEEE. 10.1109/AINA.2012.118

Zhang, H., Cheng, N., Zhang, Y., & Li, Z. (2021). Label flipping attacks against Naive Bayes on spam filtering systems. *Applied Intelligence*, *51*(7), 4503–4514. doi:10.100710489-020-02086-4

Zhang, H., Cheng, N., Zhang, Y., & Li, Z. (2021). Label flipping attacks against Naive Bayes on spam filtering systems. *Applied Intelligence*, *51*(7), 4503–4514. doi:10.100710489-020-02086-4

KEYS TERMS AND DEFINITIONS

Bioinspiration: is a paradigm shift that leads designers to draw inspiration from nature to develop new systems. Bioinspiration often relies on biomimicry.

Healthcare Email: email marketing for healthcare is about driving patient engagement.

Machine Learning: is a form of artificial intelligence (AI) that focuses on creating systems that learn, or improve their performance, based on the data they process.

Medical Event Definition: A medical event is an occurrence that results in substantial, costly, medical expenses.

Chapter 8
Artificial Intelligence in the Healthcare Sector

Zadi Salah Eddine
Dr. Moulay Tahar University, Algeria

ABSTRACT

This chapter discusses different aspect of AI in healthcare like capabilities, challenges and what values it brings to the table. It indicates that the AI is a reflexive field and can adopt with different situations and new cases easily, also making it available for everyone, anywhere, any time. It brings real world solutions because of the technical capability in different areas of foundation and specialization, like deep learning, bio-robotics, natural language proccessing. It is available and expected to be enhanced in the future, allowing it to have a bright future.

INTRODUCTION

AI in Health Care

This chapter discusses and opens doors for discussing the trending areas of technical capability of AI in the healthcare sector in general, and its areas of specialization, giving a specific example of solving a prediction problem at the end of the chapter.

It might be irregular for a layman to trust and feel secure when he is in a hospital getting treated by a machine instead of a doctor, on the other hand, it's one of the ultimate goals the AI sect is trying to achieve.

The personalized patient examination is very important in healthcare. To do so effectively, you need to be precise in identifying the illness in a reasonable time. In point of fact, this is considered an advanced type of healthcare, and it's reserved

DOI: 10.4018/978-1-6684-5656-9.ch008

for plutocratic class only. However, predictive analysis and AI could allow the democratization with a reasonable charge. Combining the historical data gathers by IoT devices, a doctor's experience, and a data scientist, will lead to gain insights from the patient's entire medical history. And hopefully extract hidden patterns also predict accurately what: the next step will be (treatment, pharmaceutical, etc.), or leads to (injury, Complications, recovery, etc.) (Prabu 2021).

CAPABILITIES OF AI IN HEALTHCARE

Supportive Tasks

Recording notes and reviewing medical records in electronic health records takes up 34% to 55% of physicians' time AI have the ability to automate the process and recuperate that percentage back to the physician's productivity by using NLP kit as a clinical documentation tool (Steven et al,. 2017), this will give more time to focus on delivering top quality care. (Natalie, Octobre 2021)

The supportive tasks can be either directed to the organization or to the patient himself.

Chat Bots

A chatbot is an artificial intelligence (AI) program that simulates and interprets human communication (written or spoken) with a purpose.

Button based chat bots basically requires little professional expertise and designed for small process automation activity with clearly defined communication scenarios like onboarding and surveying. As for the keyword based chat bots, they interact depending on the user's input(s). This will give a better experience and feels like you are texting someone, and he replies to that specific text. Finally, the contextual chat bots are more complex because it must be well learned and learnable while having the conversation and can handle the context properly and the conversation as a whole. (Ria Jaya.S, March 2019)

Progressive Treatment

The algorithm will help indicate the response of the patient towards the treatment(s) by cross referencing similar cases and comparing their treatments and outcomes.

BOOSTING PATIENT ENGAGEMENT

Prescription adherence can affect heavily on the outcome, but sometimes the patient himself fails to adjust his behavior, especially when dealing with mental health cases. On the other hand, we have the traditional care plan which will probably fail

AI can involve the patient directly into the engagement, using alerts or content intended to provoke action (Agrawal & Stethee 2018).

Improve Patient's Examination

ML strength lies in clustering, classification, prediction, anomaly detection, ranking, and automation. This can be translated to the healthcare sect in detecting similar cases, predicting and evaluating patient health, suggesting medicines or information even without asking for it, detecting rare cases, handling repetitive tasks like: data entry, scheduling, and so on.

Precision Medicine

Its foundation relies on the large amounts of data gathered from several disruptive technological innovations, such as cheap genome sequencing, advanced biotechnology, and health sensors used by patients at home and banks on advanced supercomputing algorithms with deep learning. Intel has developed a deep learning algorithm. It could identify 23 patients with high cardiovascular disease risk, undetected by traditional statistic methods.

HOW AI WILL IMPACT HEALTHCARE

Financially

Apart from saving people's life, it's observable that AI wearables and other utilities used in the health care sect will require a workforce and by that companies will compete to offer the best quality services to attract the potential patient and make it reasonable for him to pay for it, automating repetitive tasks will give more time to the whole staff involved in healthcare sect to be productive, when the service is high quality, affordable and save time it will save huge amount of money and millions of traditional work hours will be freed up (Eliana et al., 2020).

Medical Errors Minimization

An area with high population and growth rate is a challenge in terms of decent healthcare facilities. This can lead to medical errors, and AI is made to reduce them using:

- Clinical procedure's optimization. Reducing some effort and work hours from the doctor's shoulders can lead to accurate decisions. Unlike humans, AI requires no sleep and can work under stress 24/7. This will reduce human's unintentional faults and can be trustworthy.
- Communication with chat bots is a straight forward strategy to inquire patient's medical details. High quality information repertoires can find connections between symptoms and causes and minimize errors.
- Assistants in AI can arrange live video consultations anywhere any time, so there will be no need for waiting in queues for hours to meet a doctor.
- AI has the ability to analyze images without making unintentional errors, and also advanced models are more capable of discovering patterns in large amount of data in just seconds, and unlike humans AI will be more efficient with the passage of time.
- Sometimes analyzing the patient's sleep with traditional methods can be difficult, time consuming, and have wrong results; with the use of AI the process will be automated to detect sleep disorders and eventually discover the complex patterns doctors are looking for (Arindam, 2017).

Challenges Against AI In Healthcare

Deep learning projects in healthcare need to solve quite a few issues before AI can have complete domination over traditional technologies (Joe.T, 2021).

Retrospective Versus Prospective Model Training

Machine learning applications usually train their models based on historical data. When switched to real world data, they tend to output worse results, simply because the training and real-world data differ. Wearable technology, when applicable, can be the tool that helps ML algorithms adjust to production settings.

Metrics Don't Translate into Clinical Applicability

As no single measure fully describes all the properties of a machine learning model, we must rely on several measures to summarize a model's performance. However, none of these measures point out what's most important to patients, namely whether using the model leads to a positive change in patient care.

Datasets Shift

Predictive algorithms applied in a healthcare practice don't exist in a vacuum. They need to account for changes in the settings (shifting patient populations and evolving operational practices) and the changes introduced by the ML algorithms themselves.

LOGISTICAL ISSUES IN IMPLEMENTING AI SOLUTIONS

This challenge exists due to the siloed and inconsistent nature of health data that's often distributed among several standalone platforms. Therefore, aggregating and cleansing health data for machine learning is a must have attribute of any medical M L project.

Limited Algorithmic Interpretability

One of the pain points of implementing deep learning in the medical field is the inability to interpret the decision-making process in an understandable way by healthcare professionals.

For example, machine learning algorithms used in healthcare data analytics may highlight certain trends but can't explicitly trace how they arrived at these conclusions.

Ethics of Machine Learning Algorithms Used in Healthcare

Besides the technical complexity, building a healthcare ML application also involves tackling several ethical challenges:

Privacy, fairness, transparency and conflicts of interest.

The privacy challenge should be effectively addressed on the regulatory level by complying with the H I P A A rules (Health Insurance Portability and Accountability Act of 1996) and securing patients' PHI (protected health information).

Different Areas Where Health Care Uses AI

The raw materials of AI are the areas of foundation, and the manufactured materials are areas of specialization; this chapter will speak about some various well-known areas.

Areas of Foundation

"You can have data without information, but you cannot have information without data," (Moran, n.d.). Areas of foundation are the engine of AI and its fuel is data.

Fuzzy and Computational Intelligence

First appearing in 1965, it was proposed by Lotfi Zadeh with the title "Fuzzy Sets." This field stems from the ordinary logic (true and false) in math studies—fuzzy logic adds some subjectivity to the absolute truth and making it vague or imprecise between zero and one, rather than just either true or false, to mimic the human brain when making decisions. Fuzzy logic plays a big role in developing artificial intelligence solutions and still making progress towards different sectors, and its prime is in financial services, which can be adopted to any complex problem-solving use case that suggestions, recommendations are involved in it.

As for computing intelligence, it's also called soft computing and while there is still no globalized definition of it, the title refers to the learn ability of the computer from either data or experimental observation to deploy a biologically or linguistically paradigm with respect to its theory, design, and development.

Advantages of Using Fuzzy Logic

- Sometimes classical logic is dry and unlikely to reflect on real world problems, unlike fuzzy logic.
- Fuzzy logic requires less hardware requirements compared to classical logic.
- Fuzzy logic algorithms can bring accurate and valid results with ambiguous or noisy data.

Disadvantages of Using Fuzzy Logic

- Fuzzy logic algorithms need to be validated extensively and verified on a wide range.
- Fuzzy logic always needs the supervision of experts.

Nature-Inspired Algorithms

The main goal of this field is finding inexact solutions "Heuristics" for a wide range of problems that will be impossible to solve exactly in a reasonable time (exponential growth in memory and calculation resources), this type of problems is called NP-complete problems, most common NP-complete problems are (knapsack problem, travelling salesman problem, subset sum problem…and much more).nature-inspired algorithms goal is to solve this type of problems even if they are not fully solved, but at least there is a high accuracy result in a reasonable time (requires less time/ memory resources).this type of algorithms is the approach for what exact solutions are incapable of, until now, Computational intelligence is where we find most of nature-inspired algorithms or what is called evolutionary computation.

The evolutionary computation is in three parts: (Evolutionary algorithms, Swarm intelligence, Meta Heuristics) and three of them are designed to search good solutions either by finding new algorithms or enhancing already existing algorithms with controlling it's parameters or hybridizing the whole algorithm.

Data Mining and Knowledge Discovery (KDD)

Is the science of extracting hidden patterns and implicit informations from sets of explicit data and turning raw data into knowledge, it's an interdisciplinary field (Algorithmic, Databases, Statistics, Machine Learning), it's main purpose is to automate the process of acquiring knowledge without depending on human experts and knowledge engineers by replacing them with data-mining tools and a data resource.

Knowledge Representation and Reasoning (K R R, K R & R, K R^2)

The field incorporates findings in logic and combined with the scientific facts from psychology and cognitive science about the knowledge representation and reasoning in the human brain in order to make complex systems like semantic nets, theorem provers and ontologies with the simplest formalism

MACHINE LEARNING

Is the factory of building models that most of the areas of foundation relies on (Supriya & Vijay 2021). Ml is mainly in four types:

- Supervised, which needs pre labeled data to learn from and be able to solve predictions when limited data labels are pre known and solve regression problems when data labels are not known and can be unlimited or continuous outcome.
- Unsupervised (Self Supervised) the model meant to be built will depend only on the underlying structure of data to predict outcomes; no labeled data needed.

- Semi-Supervised comes between supervised and self supervised because it needs partially labeled data; just a hint to the model is enough.
- Reinforcement, here it's a different approach with new concepts like agent, reward, and environment. Allowing the agent to alter its behavior to reach a specific goal optimally in a loop of try-errors.

Natural Language Processing (N L P):

Draws from multiple disciplines: statistics, computational linguistics, and artificial intelligence. Its main purpose is to make computers formally understand human's language

As a result, NLP can make computers communicate with humans and much more (auto-translation, converting text to speech and vice versa, auto-complete, auto-correct, generate text summary, extract patterns from text, etc.).

Deep Learning (D L)

Uses a more complex structure than machine learning algorithms and mimics how the neural networks in the human brain work; it's used when machine learning models tend to struggle because of the huge amount of data

Computer Vision (CV), Pattern Recognition

Making computers understand images and videos formally is the interdisciplinary field of Computer Vision, to automate the tasks of identifying, segmenting and altering object, as for pattern recognition systems can detect patterns accurately in a reasonable time and it can be included when dealing with any type of data: images, signals, videos, text...

Data Science

Data scientists must combine a range of skills dealing with data, statistics, manipulating data with different AI approaches using: supervised learning, deep learning, and the ability to analyze results, in order to draw informed conclusions for business leaders.

Human Machine Interaction (HMI) and Intelligent Systems

Refers to the relationship between the human user and what he sees (Holzinger et al., 2017). HMI is for making the relationship more natural, including gestures have made the communication between the human user and the machine more intuitive.

Social Acceptability and Adoption of AI

Refers to the studies of the AI's effect on the community, giving an evaluation of things to aim for and things to avoid from a non-biased third person perspective, results of studies have shown that the adoption of IA globally is increasing. (Chew 2022; Marija 2020).

AREAS OF SPECIALIZATION

Predicting Illness and Patient Outcome

It revolves around automating the process of diagnosing the patient's health status using AI tools and predictive analysis e.g., COVID-19 (Anjan et al., 2021). The auto-prediction can lead to the auto-treatment of the patient or help physicians for better decision making. (Mayorga et al., 2019).

Healthcare and Wellbeing

This field focuses on providing users with better feed-back, guidance and support for staying healthy; results have shown that the expansion of AI exhibits greater economic growth. (Makridis et al., 2020)

Bio-Robotics

It's an interdisciplinary field combines biomedical engineering, robotics, and cybernetics, the science of communications and automatic control systems in both machines and living things, for the purpose of creating machines that mimics the biological systems with the use of both biology and mechanics.

Genetic Engineering

It's a branch of bio-robotics; it uses the advantage of technology for the sake of altering biological organisms.

Bionics

It's a branch of Bio-robotics and a medical engineering field. Bionic is a combined word of biology and electronics. Bionics history goes far back to ancient Egypt.

Robotics for Health

Nowadays robots are included in the operating rooms and in the clinical settings to support the workers for the sake of streamlined workflow, risk reduction and limiting contact between persons especially during times of infectious disease outbreaks like covid-19 pandemic, providing robots offers safe environment for everyone and efficient processes that requires time and effort.

Robots for Surgical-Aid

These kinds of robots achieve high levels of accuracy in a reasonable time in complex surgeries; some of the surgical robots are fully autonomous, and there are two types of surgeries that involve robotics: Minimally invasive surgeries for the torso and Orthopedic surgeries.

Neuro-Inspired AI

It is a science based on the principle of the ability to improve AI algorithms using biological neural systems. There is evidence that shows the results of logic-based and conventional mathematics can be reached using this principle. Relied on tools like: statistics, experimental neuroscience, machine learning, computer vision, neuromorphic computing. (Neuro-inspired AI UTSA AI, n.d)

Bioinformatics

Bioinformatics combines biology, computer science, physics, chemistry, statistics and mathematical equations to develop high quality software tools to understand, analyze and interpret biological data, common uses of bioinformatics are single nucleotide polymorphisms SNPs and candidates gene identification

AI for Social Good

Nowadays AI based companies number is increasing and the technical capability is also increasing simultaneously, in the other side some technical capabilities can be outside the ethical zone or against the social or environmental laws, this is one of the main reasons that field has appeared, AI for social goods bring barriers to the AI sect for the sake of increasing social and environmental safety.

Example of Using AI to Solve a Health-Care Problem

The spectrophotometer device was made to measure the chemical make-up of Materials as a blood assay, with out it, it would be very difficult and not very accurate to measure the chemicals in the blood, and we can find this device in almost any clinical or biological lab.

A recent 2022 assay called IGRA Interferom Gamma Release Assay is used. If the IGRA is positive, it means that the patient has a tuberculosis infection. IGRA is better than the traditional methods. because it can prevent the spreading of the infection, lower the price of treatment, and be detected before the symptoms appear. AI can fit into this scenario to predict the infection instead of the doctor or lab staff which will save them time and energy to be more productive and avoid human errors. This will require the collaboration of different labs on the IGRA blood assay, spectrophotometry, automation, and machine learning developers.

The first step will be to collect enough clean data, which can be the longest phase. The next step is to integrate a smart chip (Raspberry Pi) into the device that provides the blood components.mesurement. The chip contains a predictive model that uses data provided by the device. The predictive model, when linked to it, will be able to detect the early stages of tuberculosis development, it will save doctors time and energy analyzing and reading multiple blood tests for tuberculosis germs and assay results. A machine learning approach is the best match for this case. The model will be able to predict whether the patient is infected and requires treatment or is negative based on historical data. As a future vision, other infections Related models can be integrated alongside the first model on the same device. And it will be used in airports, clinics, suspicious places.

Using AI can make the world a better place to live in.

CONCLUSION

This chapter discussess different aspectst of AI in HealthCare.Various techniques using AI have greatly helped to curb the COVID-19 outbreak by assisting with early diagnosis. This indicates that AI is a flexible field and can adapt to different situations and new cases easily because of the technical capability available and expected to be enhanced in the future and the variety of areas of research, allowing AI to have a bright future.

NOTE

Using the example of a person X who is well-versed in science A but lacking in science B, the fact that science A and B are linked indicates that the person X will lack knowledge in his A field from the door of the B science. AI with its core theories requires a certain level of math, statistics, and programming to make it real world applicable.

This chapter is meant to sharpen your non-technical skills and make you see the big picture and the fusion between AI and the healthcare sector, covering some of the advanced topics and where the bioinspiration methods are used, more technical and related academic areas like (generative adversarial network (GAN), genetic algorithms) will be covered

REFERENCES

Agrawal, J. S. (2018). an AI Powered Electronic Stethoscope. *Anaesthesia, Pain & Intensive Care*, *22*(3), 412–413.

Biundo, E., Pease, A., Segers, K., & de Groote, M. (2020). The socio-economic impact of AI in healthcare. *Deloitte*.

Chatterjee, A. (2017) Use of Artificial Intelligence to reduce Medical Errors. *Health Vectors*. https://blog.myhealthvectors.com.

Chew, H. S. J., & Achananuparp, P. (2022). Perceptions and Needs of Artificial Intelligence in Health Care to Increase Adoption: Scoping Review. *Journal of medical Internet research*, *24*(1), e32939. https://doi.org/10.2196/32939.

Cubric, M. (2020) Drivers, Barriers and social considerations for AI adoption in business and management: A tertiary study. *Technology in Society, 62*.

Gudigar, A., Raghavendra, U., Nayak, S., Ooi, C. P., Chan, W. Y., Gangavarapu, M. R., Dharmik, C., Samanth, J., Kadri, N. A., Hasikin, K., Barua, P. D., Chakraborty, S., Ciaccio, E. J., & Acharya, U. R. (2021). Role of Artificial Intelligence in COVID-19 Detection. *Sensors, 21*(23), 8045. https://doi.org/10.3390/s21238045.

Holzinger, A., Biemann, C., Pattichis, C. S., & Kell, D. B. (2017). What do we need to build explainable AI systems for the medical domain?

Jaya, R. S(2019) 5 chatbot ideas business cant afford to miss in 2019. *Hackernoon*. https://hackernoon.com/5-chatbots-ideas-business-cant-afford-to-miss-in-2019-e59cd41e01f8?source=rss----3a8144eabfe3---4

Joe, T. (2021). Deep Learning and ML in Health Care: Use Cases, Examples. *TopFlight*. https://topflightapps.com/ideas/machine-learning-in-healthcare-use-cases/

Makridis, C., & Mishra, S. (2020) How Does Artificial Intelligence Raise Well-being? Evidence from Cities in the United States.: https://ssrn.com/abstract=3669348 doi:10.2139/ssrn.3669348

Matrix. (n.d.). Neuro-inspired, AI: The UTSA AI Consortium for Human Well-Being. *Matrix*. https://ai.utsa.edu/research/neuro-inspired-artificial-intelligence/

Mayorga-Ruiz, I., Jiménez-Pastor, A., Fos-Guarinos, B., López-González, R., García-Castro, F., & Alberich-Bayarri, Á. (2019). The Role of AI in Clinical Trials. In *Artificial Intelligence in Medical Imaging* (pp. 231–243). Springer. doi:10.1007/978-3-319-94878-2_16

Natalie, T. (2021) ML in Health-care: 12 Real-World Use Cases to Know. *NixUnited* https://nix-united.com/blog/machine-learning-in-healthcare-12-real-world-use-cases-to-know/

Prabu, A. (2021). SmartScope: An AI-Powered Digital Auscultation Device To Detect Cardiopulmonary Diseases.

Steven, Y. (2017). *Lin, Tait D Shanafelt, Steven M Asch*. Reimagining Clinical Documentation With Artificial Intelligence.

Supriya, M., & Chattu, V. K. (2021). A Review of Artificial Intelligence, Big Data, and Blockchain Technology Applications in Medicine and Global Health. *Big Data and Cognitive Computing.*, *5*(3), 41. doi:10.3390/bdcc5030041

Chapter 9
Genetic Algorithm and Machine Learning

Radha Raman Chandan
https://orcid.org/0000-0001-7382-6143
School of Management Sciences, India

Sarita Soni
https://orcid.org/0000-0001-9193-1545
Government Polytechnic, Bighapur, India

Atul Raj
Government Polytechnic, Sikandra, India

Vivek Veeraiah
Adichunchanagiri University, India

Dharmesh Dhabliya
Vishwakarma Institute of Information Technology, India

Sabyasachi Pramanik
https://orcid.org/0000-0002-9431-8751
Haldia Institute of Technology, India

Ankur Gupta
https://orcid.org/0000-0002-4651-5830
Vaish College of Engineering, India

ABSTRACT

Genetic algorithm is based on the natural search process, which mimics natural growth and employs approaches inspired by natural evolution to solve optimization problems, employing bequest, mutation, and miscellany, as well as intersect. Its actual meaning is a competent, concurrent, and universal search approach that continuously obtains and builds up knowledge about search space and command management search space in order to alter the best search result. The traditional multilevel association rules mining techniques generate a large number of candidate items and compare them to the whole database. Nonetheless, the majority of mining procedures are in vain, since they guide crucial costs associated with computing. The inherited algorithms provide a novel technique for tackling these sorts of problems.

DOI: 10.4018/978-1-6684-5656-9.ch009

INTRODUCTION

John Holland created the Genetic Algorithm (Katoch et al., 2021) (GA) in the year 1970. It is based on a genetic organism's hereditary mechanism. Normal humans developed through many generations according to the idea of normal taxonomy and the "survival of the fittest," according to Charles Darwin, who claimed that the genesis of living things was undeniable. GA is an adaptive procedure that is used to tackle problems involving search and optimization (Hashim et al., 2021). No more explanation can be found after a large number of new generations have been created using the outlined technique. This answer is regarded as the ultimate outcome.

The vast amount of data we have access to can be separated into tiny groupings, each of which may be assessed as a population. Re-applying hereditary operators to the populace is the most beneficial approach to the present predicament. As we all know, search progress is a problem-solving process in which we can't predict the sequence of actions that will lead to interpretation in future publications. Based on how well and smartly we employed search operators to accomplish this goal. A good search mechanism should be capable of doing searches both locally and in a random manner. Local search investigates all local capabilities and helps to reach the best solution as far as possible, while random search (Wu et al., 2021) explores the whole solution and is effective at avoiding the most favourable local.

Figure 1. Genetic algorithm flow chart

{

1. [Start] Generate random population of n chromosomes (suitable solutions for the problem).
2. [Fitness] Evaluate the fitness f(x) of each chromosome x in the population.
3. [New population] Create a new population by repeating following steps until the new population is complete
   a. [Selection] Select two parent chromosomes from a population according to their fitness (the better fitness, the bigger chance to be selected)
   b. [Crossover] With a crossover probability cross over the parents to form a new offspring (children). If no crossover was performed, offspring is an exact copy of parents.
   c. [Mutation] With a mutation probability mutate new offspring at each locus (position inchromosome).
   d. [Accepting] Place new offspring in a new population.
4. [Replace] Use new generated population for a further run of algorithm.
5. [Test] If the end condition is satisfied, stop, and return the best solution in current population.
6. [Loop] Go to step 2.

}

Procedure

1. Use the Genetic K-Means Algorithm

We may have an algorithm that combines the progress of the genetic algorithm with the K-Means method for clustering, in addition to parallel implementation utilising genetic algorithms. Optimal clustering (Pramanik et al., 2020) is more natural than K-Means clustering, however it typically comes with certain short-term drawbacks.

ANALYSIS OF PRIMARY COMPONENTS (PCA)

In this case, the data dimension is given additional weight in order to facilitate computation. Let's examine at two-dimensional data to see how PCA (Pramanik et al., 2021) works. When data is plotted on a graph, two axes are created. PCA works with data, which is subsequently transformed into a single dimension. This is shown in Figure 1, and PCA pseudo-code is presented in Figure 2.

Figure 2. Shows data visualisation before and after PCA

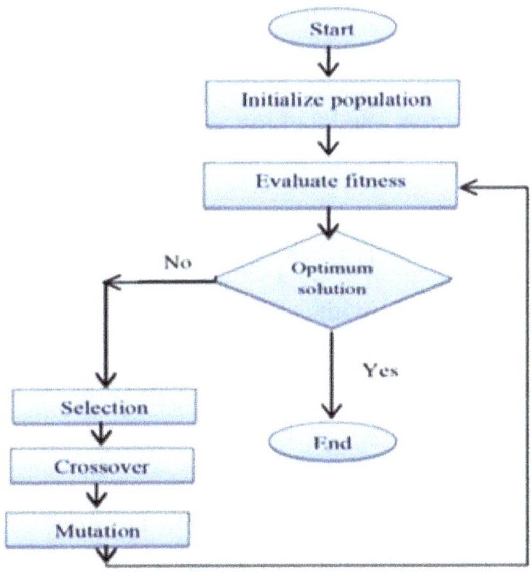

Figure 3. Pseudocode for PCA

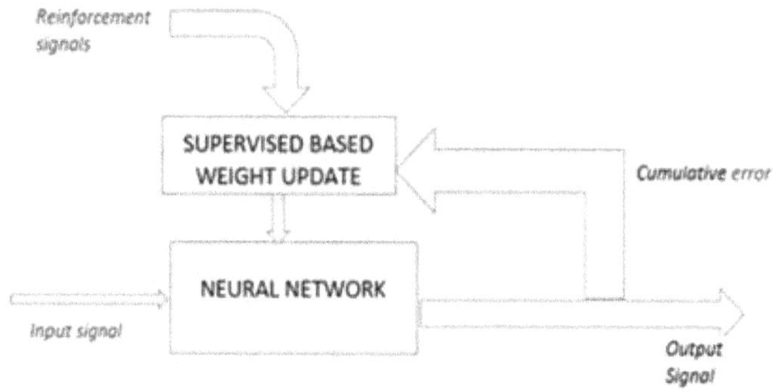

Learning in a Semi-Supervised Environment

It incorporates both supervised and unsupervised learning (Dutta et al., 2021) strategies. In machine learning and data mining (Samanta et al., 2021), unlabeled data may be more useful than labelled data, yet effective operations on labelled data are very challenging. Semi-supervisory learning is divided into numerous categories. The following are some examples:

Generative Models are a kind of model that is used to generate new ideas. p (x, y) = p (y) p (x I y) p (x I y) p (x I y) p (x I y) p (x I y) p (x I y) p (x I y) p (x I y) p (x I y) p (x I y) p (The Gaussian mixture model is a kind of mathematical model. Through unlabeled data that might be personal, mixed machines are created. To assure mixture dispersion, just one label instance per ingredient is required.

Self-Training

A classifier qualifies a collection of data with a label in this case. The classifier is then given unlabeled data to work with. After that, the unlabeled point and anticipated label are combined with the prepared set. Since then, the system has remained steady. Self-training gets its name from the ability to categorise oneself.

Transductive SVM

Transductive Support Vector Machine (TSVM) (Dushyant et al., 2022) is a kind of support vector machine that incorporates SVM. It is feasible to measure both labelled and unlabeled data with TSVM. The TSVM algorithm is used to convert unlabeled data to label data, which has the largest margin between label and unlabeled data. Using TSVM to find the proper answer is an NP-hard task.

i) Learning while multitasking

The goal of multiple learning is to assist prior learners in improving their performance. When a task is completed, the Multi-task Learning algorithm is activated; it remembers how to solve the issue or finish the job. These stages are used by the algorithm to address a similar issue or job. The inductive transfer mechanism is a supplementary function provided by this method. When learners share their own expertise with one another, they may learn far more rapidly and without feeling lonely.

ii) Ensemble Learning

Ensemble learning, such as Naive Bayes (Kaushik et al., 2021), Decision Tree (Bhattacharya et al., 2021), and Neural Network (Meslie et al., 2021), is a kind of learning in which individual learners interact to form a single learner. Ensemble learning is a relatively new field that dates back to the 1990s. Individual learners may have been approximately improved for eternity, but the group of learners working faster than individual learners may have been roughly improved for eternity. Two widely used Ensemble learning approaches are as follows:

1) Boosting: Boosting is a strategy for reducing bias and inconsistency in assembly learning. Boosting is the process of taking a group of weak learners and replacing one of them with a strong learner. A weak learner is a classifier that does not have any real-world associations. A strapping learner, on the other hand, is a classifier with a high correlation in real classification. AdaBoost (Pramanik et. al., 2022) is the most well-known example of boosting, and the pseudocode of AdaBoost is given in Figure 3.

Figure 4. AdaBoost pseudocode

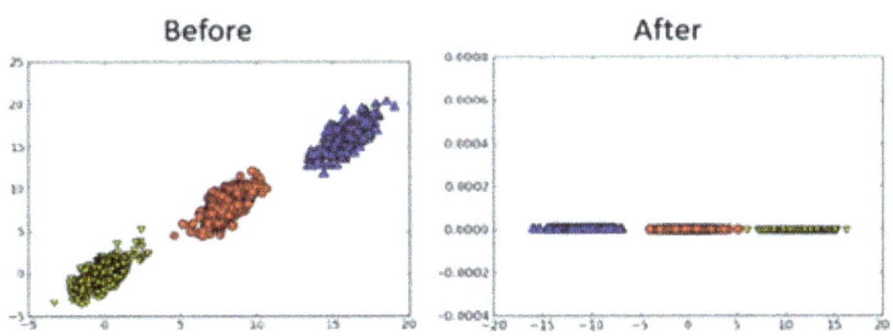

2) Bagging: A bagging or bootstrap aggregate function is one in which the accuracy and consistency requirements for a machine learning method are big enough. This is important when it comes to categorization and regression. Bags are useful for eliminating discrepancies and assisting with treatment overfitting. Figure 4 shows the pseudocode for bagging.

Figure 5. Bagging Pseudocode

```
R ← X
for(k = 0, ..., K − 1) do
    {
    λ = 0
    T^(k) ← R^(k)
    for(j = 0, ..., J) do
        {
        P^(k) ← R^T T^(k)
        P^(k) ← P^(k) ||P^(k)||^−1
        T^(k) ← RP^(k)
        λ' ← ||T^(k)||
        if(|λ' − λ| ≤ ε) then break
        λ ← λ'
        }
    R ← R − T^(k)(P^(k))^T
    }
return T, P, R
```

iii. Neural Network Learning

The neural network, also known as an Artificial Neural Network (ANN) (Bacanin et al., 2022), is not derived from organic neurons. Inside the brain, a neuron is a single cell with a similar makeup. To comprehend a neural network, one must first comprehend the function of a neuron. Dendrite, nucleus and axon are the four major components of a neuron, as depicted in Fig. 5.

Electrical signals are received by dendrites. Electrical signals are processed by Soma. Soma's output travels via an axon to the terminal of a dendrite. This terminal serves as an interface, allowing the output of this neuron to be transmitted to the next neuron. Nucleus is the most sensitive portion of a neuron. The neural network, which is found in the brain and where electrical impulses travel about, is made up of interconnected neuron.

ANN functions in the same way as a neural network. There are three levels to it. The input layer (similar to a dendrite) is in charge of receiving data, while the hidden layer (Mandal et al., 2021) is in charge of processing it (like soma and axon).

Figure 6. A Neuron

Input: Data set $\mathcal{D} = \{(\boldsymbol{x}_1, y_1), (\boldsymbol{x}_2, y_2), \cdots, (\boldsymbol{x}_m, y_m)\}$;
Base learning algorithm $\mathcal{L}$;
Number of learning rounds T.

Process:

$D_1(i) = 1/m$.

for $t = 1, \cdots, T$:

$h_t = \mathcal{L}(\mathcal{D}, D_t)$;

$\epsilon_t = \Pr_{i \sim D_t}[h_t(\boldsymbol{x}_i \neq y_i)]$;

$\alpha_t = \frac{1}{2}\ln \frac{1-\epsilon_t}{\epsilon_t}$;

$$D_{t+1}(i) = \frac{D_t(i)}{Z_t} \times \begin{matrix} \exp(-\alpha_t) & \text{if } h_t(\boldsymbol{x}_i) = y_i \\ \exp(\alpha_t) & \text{if } h_t(\boldsymbol{x}_i) \neq y_i \end{matrix} = \frac{D_t(i)\exp(-\alpha_t y_i h_t(\boldsymbol{x}_i))}{Z_t}$$

end.

Output: $H(\boldsymbol{x}) = \text{sign}\,(f\,(\boldsymbol{x})) = \text{sign} \;\sum_{t=1}^{T} \alpha_t h_t(\boldsymbol{x})$

Figure 7. Artificial Neural Network Structure

Input: Data set $\mathcal{D} = \{(\boldsymbol{x}_1, y_1), (\boldsymbol{x}_2, y_2), \cdots, (\boldsymbol{x}_m, y_m)\}$;
Base learning algorithm $\mathcal{L}$;
Number of learning rounds T.

Process:

for $t = 1, \cdots, T$:

$\mathcal{D}_t = Bootstrap(\mathcal{D})$;

$h_t = \mathcal{L}(\mathcal{D}_t)$

end.

Output: $H(\boldsymbol{x}) = \text{argmax}_{y \in \mathcal{Y}} \sum_{t=1}^{T} 1(y = h_t\,(\boldsymbol{x}))$

Finally, the output layer is in charge of transmitting processed output to dendritic terminals. There are three different types of Artificial Neural Networks: supervise, unsupervised, and reinforce.

i) Supervised Neural Network

The output of a Supervised Neural Network is already known about the provided input. In this case, the predicted outputs were compared to the actual outcome. The number of parameters (Ahmad et al., 2021) is modified based on the output error and then injected into the neural network. Figure 7 depicts the overall development. As a result, as a feed forward neural network, a supervised neural network is used.

ii) Unsupervised Neural Network

There is no suggestion for output in an Unsupervised Neural Network for a specific input. This network's most significant role is to prepare input data based on a set of comparable characteristics. The neural network now checks input combinations and groups them appropriately. This network's block diagram is presented in Figure 8.

Figure 8. Supervised Neural Network

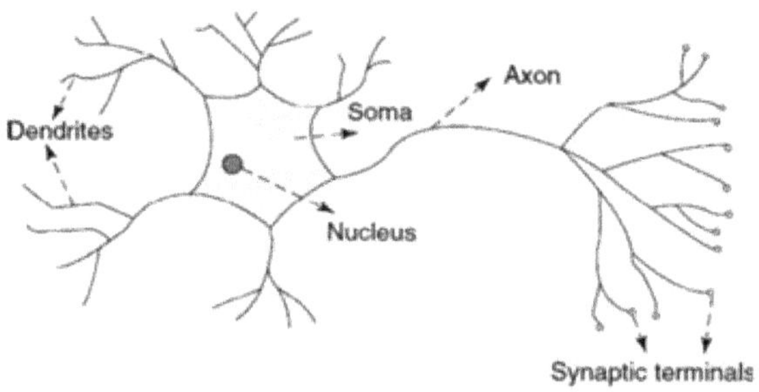

iii) Reinforced Neural Network

The network offers a mechanism for individuals to interact with others around them in a Reinforced Neural Network (Bassi and Attux, 2022). The network responds by confirming whether the choice made via the network is accurate or erroneous based on the circumstances. If the result is right, the relationships pointing to output accuracy will be strengthened. The organisation is in a state of flux. Aside from that, there is no previous information on the network's availability. Figure 3.19 depicts the reinforced neural network.

Figure 9. Unsupervised neural network

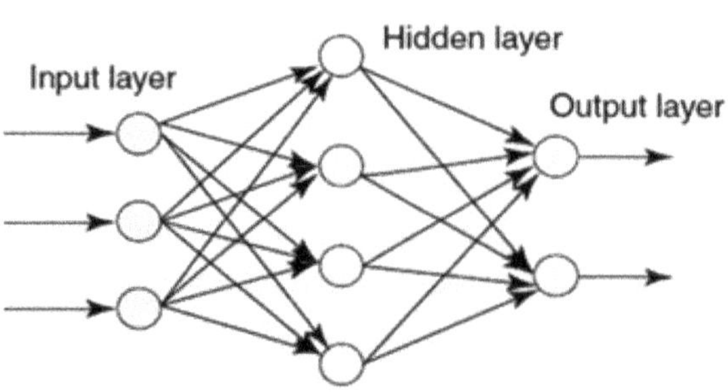

iv) Instance-Based Learning

Instance-Based learning is a precise model for a learner's learning. It tries to connect the corresponding form to data that was previously supplied. As a result, the most fitting word for this is instance-based. It is utilised by a lazy learner who is waiting for test data to arrive before interacting with the training data. As the bulk of the data becomes larger, the complexity of this knowledge-based method grows. The kernel machine and the RPF network are two well-known examples of instance-based learning that employ K-nearest neighbour.

Figure 10. Reinforced Neural Network (RNN)

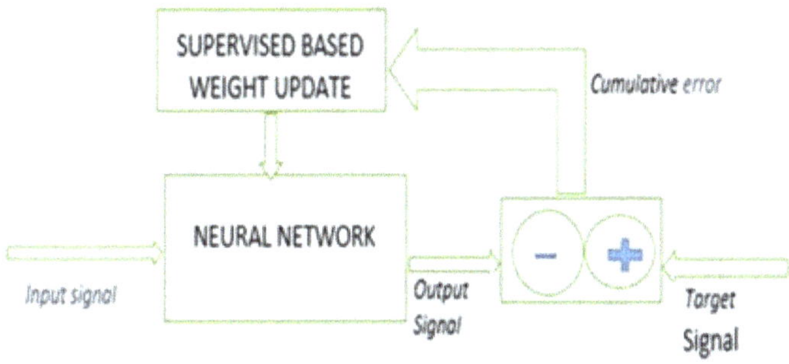

TOOLS FOR MACHINE LEARNING

Scikit-learn, PyTorch, WEKA (Sakshi et al., 2021), TensorFlow (Baeta et al., 2021), RapidMiner, RStudio (Hu, 2021), and other machine learning tools are available, and the following is a description of this method:

Scikit-learn is an Open-Source Learning Framework (Sklearn)

Python, which includes Scikit-learn (Agrawal, 2021), is utilised for many machine learning development tasks. Python is a computer language that comes with a wide collection of development tools. Scikit-source learn's code is developed in the Python programming language and is based on NumPy (Pramanik and Bandyopadhyay, 2022), SciPy (Pramanik and Bandyopadhyay 2022), and Matplotib.

It has the following features:

1. It has data mining (Pramanik et al., 2022) and data analysis capabilities.

2. It covers classification, regression, clustering, dimension reduction, model selection, and pre-processing, among other topics.

Pros:

3. The instructional material is simple to comprehend and put into practise.
4. Documentation is supplied that is simple to grasp.
5. Different parameters for every algorithm may be changed according to requirements in order to invoke a particular object

PyTorch

Facebook's AI research department has urbanised it. It is based on the PyTorch library (Imambi et al., 2021). It's an open-source machine learning library that may be used for deep learning, natural language processing, and computer vision (Jayasingh et al., 2022), among other things.

It has the following features:

1. It assists in the creation of neural networks with the Auto grade module.
2. It provides a series of optimization algorithms for structural neural networks.
3. It may be used on a cloud-based platform.
4. It also supports the use of a variety of different tools, libraries, and distributed training.

Pros:

5. It gives enough assistance in the creation of a computation map.
6. It has a variety of front ends that are simple to use.

WEKA

It is an open-source machine learning programme that has been tried and proven. It's written in Java and can run on any platform. It contains a collection of machine learning algorithms with the goal of assisting in the resolution of real-world data mining challenges.

Characteristics:

1. Data preparation
2. Classification
3. Regression

4. Clustering
5. Visualization
6. Association rules and regulations, etc.

Pros:

7. A variety of web-based courses are available to give training.
8. All algorithms are straightforward to understand and implement.
9. The application's versatility expands the student's options.

Cons:

10. There are a limited number of supporting materials and internet resources accessible.

TensorFlow

It is a free end-to-end open-source machine learning platform created by the Google Brain team. It allows for the creation of a JavaScript record, which aids in machine learning. Various APIs are utilised to train and develop the model here.

Description:
Step 1: To train and build the model.
Step 2: Using TensorFlow.js to run the model (model converter).
Pros:

1. It may be utilised by utilising character tags or installing using Node Package Manager (NPM).
2. Human stance inference is easily applied.
   Cons:
3. Learning is tough due to the difficulties in obtaining study materials.

RapidMiner

It is used for research, study, and the creation of various applications. RapidMiner provides a platform for machine learning, deep learning, data analysis, text mining, and predictive analytics, among other things.

Features:

1. It aids in the creation and implementation of logical processes in the system.
2. It helps with data training.

3. It employs the usage of result visualisation.
4. Validation and optimization of models are possible.

Pros:

5. Plug-in power that can be expanded.
6. The method of using it is simple.
7. You don't need a lot of programming knowledge.

Cons:

8. Each tool is quite expensive to use.
9. Sharing RapidMiner Studio analysis is really challenging.
10. We can only deal with 10,000 rows in the trial version.
11. The corporate team's reaction is adequate.

RStudio

The R programming language makes use of RStudio. RStudio provides the R integrated development environment (IDE), which includes a terminal and a syntax-highlighting editor. Support for graphics and statistical computation in the R programming language. RStudio Desktop and RStudio Serve are the two versions available. RStudio Desktop is a form of desktop programme that runs on a workstation with a local operating system. RStudio Server is a server-side programme that runs on a remote server and allows access to RStudio through any web browser.

RStudio Desktop and RStudio Server are available in both commercial and non-commercial variants. The format or edition of the IDE determines the OS's long-term viability. RStudio Desktop is available as a packaged base version for Windows, macOS, and Linux. Debian, Ubuntu, Red Hat Linux, CentOS, openSUSE, and SLES are all supported by RStudio Server and Server Pro.

Open-source software development is one of the advantages of R programming.

2. Lacking graphics prowess.
3. A society that is very active.
4. A large number of packages are available.
5. An all-encompassing atmosphere
6. Compute the composite statistical data.
7. Encourage the use of distributed computing.
8. Run code without using a compiler.

Pros:

9. Communicative.
10. Debug with little effort.
11. Feature of auto-completion.
12. A sociable atmosphere in which to build packages.
13. Codes are auto-saved, and work is organised as a project.

Cons:

14. It often causes viewers to have a crisis.
15. Inability to see all fields at the same time (due to HTML boundary).

CONCLUSION

The genetic algorithm is based on the natural search process, which imitates natural growth. Its true meaning is a competent, concurrent, and universal search method that continually gathers and produces information about the command management search space and the search space for searches in order to change the best search result. The conventional methods for mining multilevel association rules produce a huge number of candidate items and evaluate them against the whole database. The bulk of mining techniques, however, are useless since they influence significant computational costs. The inherited algorithms provide a fresh approach to solving these kinds of issues. The GA-dependent technique efficiently tests the common predicted main aspirant items to control the search space and arrive at the best answer while searching for association rules. Due to this alluring advantage, association laws may considerably restrict the right to conduct searches while enhancing mining operations.

REFERENCES

Agrawal, T. (2021). Hyperparameter Optimization Using Scikit-Learn. In *Hyperparameter Optimization in Machine Learning*. Apress. doi:10.1007/978-1-4842-6579-6_2

Ahmad, A., Garhwal, S., Ray, S. K., Kumar, G., Malebary, S. J., & Barukab, O. M. (2021). The Number of Confirmed Cases of Covid-19 by using Machine Learning: Methods and Challenges. *Archives of Computational Methods in Engineering*, *28*(4), 2645–2653. doi:10.100711831-020-09472-8 PMID:32837183

Bacanin, N., Vukobrat, N., Zivkovic, M., Bezdan, T., & Strumberger, I. (2022). Improved Harris Hawks Optimization Adapted for Artificial Neural Network Training. In C. Kahraman, S. Cebi, S. Cevik Onar, B. Oztaysi, A. C. Tolga, & I. U. Sari (Eds.), *Intelligent and Fuzzy Techniques for Emerging Conditions and Digital Transformation. INFUS 2021. Lecture Notes in Networks and Systems* (Vol. 308). Springer. doi:10.1007/978-3-030-85577-2_33

Baeta, F., Correia, J., Martins, T., & Machado, P. (2021). TensorGP – Genetic Programming Engine in TensorFlow. In P. A. Castillo & J. L. Jiménez Laredo (Eds.), Lecture Notes in Computer Science Vol. 12694. Springer. doi:10.1007/978-3-030-72699-7_48

Bassi, P. R. A. S., & Attux, R. (2022). A deep convolutional neural network for COVID-19 detection using chest X-rays. *Research on Biomedical Engineering*, *38*(1), 139–148. doi:10.100742600-021-00132-9

Bhattacharya, A., Ghosal, A., Obaid, A. J., Krit, S., Shukla, V. K., Mandal, K., & Pramanik, S. (2021). Unsupervised Summarization Approach with Computational Statistics of Microblog Data. In D. Samanta, R. R. Althar, S. Pramanik, & S. Dutta (Eds.), *Methodologies and Applications of Computational Statistics for Machine Learning* (pp. 23–37). IGI Global. doi:10.4018/978-1-7998-7701-1.ch002

Dushyant, K., Muskan, G., Gupta, A., & Pramanik, S. (2022). Utilizing Machine Learning and Deep Learning in Cyber security: An Innovative Approach. In M. M. Ghonge, S. Pramanik, R. Mangrulkar, & D. N. Le (Eds.), *Cyber security and Digital Forensics*. Wiley. doi:10.1002/9781119795667.ch12

Dutta, S., Pramanik, S., & Bandyopadhyay, S. K. (2021). Prediction of Weight Gainduring COVID-19 for Avoiding Complication in Health. *International Journal of Medical Science and Current Research*, *4*(3), 1042–1052.

Hashim, F. A., Hussain, K., Houssein, E. H., Mabrouk, M. S., & Al-Atabany, W. (2021). Archimedes optimization algorithm: A new metaheuristic algorithm for solving optimization problems. *Applied Intelligence*, *51*(3), 1531–1551. doi:10.100710489-020-01893-z

Hu, K. (2021). Become Competent in Generating RNA-Seq Heat Maps in One Day for Novices Without Prior R Experience. In K. Hu (Ed.), *Nuclear Reprogramming. Methods in Molecular Biology* (Vol. 2239). Humana. doi:10.1007/978-1-0716-1084-8_17

Imambi, S., Prakash, K. B., & Kanagachidambaresan, G. R. (2021). PyTorch. In K. B. Prakash & G. R. Kanagachidambaresan (Eds.), *Programming with TensorFlow. EAI/Springer Innovations in Communication and Computing*. Springer. doi:10.1007/978-3-030-57077-4_10

Jayasingh, R., Kumar, J., Telagathoti, B., Sagayam, K. M., & Pramanik, S. (2022). Speckle noise removal by SORAMA segmentation in Digital Image Processing to facilitate precise robotic surgery. *International Journal of Reliable and Quality E-Healthcare*, *11*(1), 2022. doi:10.4018/IJRQEH.295083

Katoch, S., Chauhan, S. S., & Kumar, V. (2021). A review on genetic algorithm: Past, present, and future. *Multimedia Tools and Applications*, *80*(5), 8091–8126. doi:10.100711042-020-10139-6 PMID:33162782

Kaushik, D., & Garg, M. Annu, Gupta, A. and Pramanik, S. (2021). Application of Machine Learning and Deep Learning in Cyber security: An Innovative Approach. In M. Ghonge, S. Pramanik, R. Mangrulkar and D. N. Le, (eds.), Cybersecurity and Digital Forensics: Challenges and Future Trends. Wiley.

Mandal, A., Dutta, S., & Pramanik, S. (2021). Machine Intelligence of Pi from Geometrical Figures with Variable Parameters using SCILab. In D. Samanta, R. R. Althar, S. Pramanik, & S. Dutta (Eds.), *Methodologies and Applications of Computational Statistics for Machine Learning* (pp. 38–63). IGI Global., doi:10.4018/978-1-7998-7701-1.ch003

Meslie, Y., Enbeyle, W., Pandey, B. K., Pramanik, S., Pandey, D., Dadeech, P., Belay, A., & Saini, A. (2021). Machine Intelligence-based Trend Analysis of COVID-19 for Total Daily Confirmed Cases in Asia and Africa. In D. Samanta, R. R. Althar, S. Pramanik, & S. Dutta (Eds.), *Methodologies and Applications of Computational Statistics for Machine Learning* (pp. 164–185). IGI Global., doi:10.4018/978-1-7998-7701-1.ch009

Pramanik, S. (2022). Carpooling Solutions using Machine Learning Tools. In *Handbook of Research on Evolving Designs and Innovation in ICT and Intelligent Systems for Real-World Applications, K. K. Sarma, N. Saikia and M. Sharma*. IGI Global. doi:10.4018/978-1-7998-9795-8.ch002

Pramanik, S., & Bandyopadhyay, S. (2022). Identifying Disease and Diagnosis in Females using Machine Learning. In I. G. I. John Wang (Ed.), *Encyclopedia of Data Science and Machine Learning*. Global.

Pramanik, S., & Bandyopadhyay, S. (2022). Analysis of Big Data. In I. G. I. John Wang (Ed.), *Encyclopedia of Data Science and Machine Learning*. Global.

Pramanik, S., Galety, M. G., Samanta, D., & Joseph, N. P. (2022). Data Mining Approaches for Decision Support Systems. *3rd International Conference on Emerging Technologies in Data Mining and Information Security*.

Pramanik, S., Sagayam, K. M., & Jena, O. P. (2021). Machine Learning Frameworks in Cancer Detection. ICCSRE 2021, Morocco, North Africa.

Pramanik, S., Singh, R. P., & Ghosh, R. (2020). Application of Bi-orthogonal Wavelet Transformand Genetic Algorithm in Image Steganography. *Multimedia Tools and Applications*, *79*(25-26), 17463–17482. doi:10.100711042-020-08676-1

Sakshi, G. S., Sharma, C., Kukreja, V. (2021). Handwritten Mathematical Symbols Classification Using WEKA. In Choudhary, A., Agrawal, A.P., Logeswaran, R., Unhelkar, B. (eds) Applications of Artificial Intelligence and Machine Learning. Lecture Notes in Electrical Engineering (vol. 778). Springer, Singapore. doi:10.1007/978-981-16-3067-5_4

Samanta, D., Dutta, S., Galety, M. G., & Pramanik, S. (2021). A Novel Approach for Web Mining Taxonomy for High-Performance Computing, The *4th International Conference of Computer Science and Renewable Energies (ICCSRE'2021)*. doi:10.1051/e3sconf/202129701073

Wu, D., Liao, Y., Hu, C., Yu, S., & Tian, Q. (2021). An Enhanced Fuzzy Control Strategy for Low-Level Thrusters in Marine Dynamic Positioning Systems Based on Chaotic Random Distribution Harmony Search. *International Journal of Fuzzy Systems*, *23*(6), 1823–1839. doi:10.100740815-020-00989-5

Chapter 10
Application of Bioinformatics in Cancer Prediction and Prognosis

Bekaddour Abdel Madjid
Independent Researcher, Algeria

ABSTRACT

Informatics and internet technologies are becoming extremely popular in today's healthcare system. The emergence of the worldwide web has affected the way in which health-related information is distributed and accessed over cyberspace. The internet is rapidly gaining importance, not just for healthcare professionals, but also for patients, by enabling them to search for drug- and other health-related information. Bioinformatics is the combination of biology and information technology. The term bioinformatics was coined by Paulien Hogeweg in 1979 for the study of informatic processes in biotic systems. Its primary use since the late 1980s has been in genomics and genetics, particularly in those areas of genomics involving large-scale DNA sequencing.

INTRODUCTION

Informatics and Internet technologies have become very common in today's healthcare system. The emergence of the web around the world has affected the way health-related information is distributed and accessed through cyberspace. The Internet is rapidly gaining importance, not only for healthcare professionals, but also for patients, by enabling them to search for information health (Whitman, 2022). So in this chapter

DOI: 10.4018/978-1-6684-5656-9.ch010

we will see a definition bioinformatics and the components of bioinformatics and The value of Bioinformatics in ML and Tumor image segmentation.

Bioinformatics is the combination of biology and information technology. (Anderson, 2002). The term bioinformatics was coined by Paulien Hogeweg in 1979 for the study of informatic processes in biotic systems. It was primary used since late 1980s has been in genomics and genetics, particularly in those areas of genomics involving large-scale DNA sequencing.

Bioinformatics can be defined as the application of computer technology to the management of biological information. Bioinformatics is the science of storing, extracting, organizing, analyzing, interpreting and utilizing information from biological sequences and molecules. It has been mainly fueled by advances in DNA sequencing and mapping techniques. Over the past few decades rapid developments in genomic and other molecular research technologies and developments in information technologies have combined to produce a tremendous amount of information related to molecular biology. The primary goal of bioinformatics is to increase the understanding of biological processes. (Whitman, 2022)

The Components of Bioinformatics: The discipline encompasses any computational tools and methods used to manage, analyze and manipulate large sets of biological data in order to:

- The creation of databases, allowing the storage and management of large biological data sets
- The development of algorithms and statistics to determine relationships among members of large data sets
- The use of these tools for the analysis and interpretation of various types of biological data, including DNA, RNA and protein sequences, protein structures, gene expression profiles, and biochemical pathways (Whitman, 2022).

The Bioinformatics focuses on the development of algorithms and software for the transfer, storage, analysis, and development of genomics databases. Machine learning (ML) belongs to the branch of computer science that provides self-learning capability to the machines without explicit programming. The ML algorithms are being extensively used for the tasks of prediction, classification, and feature selection in bioinformatics. The ML approaches are very good for solving problems such as distinguishing between DNA sequences and classification of DNA sequences. Currently, the ML in bioinformatics has become significant due to the advent of deep learning (Guo, 2018).

Bioinformatics represents an interdisciplinary branch for developing improved methods for retrieving, analyzing, storing, and organizing the biological data. It focuses on the development of algorithms and software for the transfer, storage, analysis, and development of genomics database. Machine learning (ML) belongs to the branch of computer science that provides self-learning capability to the machines without explicit programming. The ML algorithms are being extensively used for the tasks of prediction, classification, and feature selection in bioinformatics. The ML approaches are very good for solving problems such as distinguishing between DNA sequences and classification of DNA sequences. Currently, the ML in bioinformatics has become significant due to the advent of deep learning (Malik, 2021)

The ML algorithms are used for selecting relevant features in bioinformatics application as biological data are high dimensional in nature (Malik, 2021). The second major application of ML in bioinformatics has been found in classification of biological data. The third major application of ML in bioinformatics was found to be prediction of biological data (Malik, 2021).

TUMOR

An abnormal mass of tissue that forms when cells grow and divide more than they should or do not die when they should. Tumors may be benign (not cancerous) or malignant (cancerous). Benign tumors may grow large but do not spread into, or invade, nearby tissues or other parts of the body. Malignant tumors can spread into, or invade, nearby tissues. They can also spread to other parts of the body through the blood and lymph systems. Also called neoplasm (Richters, 2019).

Tumor in Africa

Cancer care in Africa has greatly improved over the last 10 years. By 2017, 71% of African countries, up from 46% in 2013, had operational national cancer control programmes, and the oncology workforce has increased steadily. However, care and services vary widely, sometimes within the same country or region from institutions offering state-of-the-art care to basic services that can be hard to access. Many patients in Africa are diagnosed with advanced cancers and do not complete their care. There are several reasons for this, cost being the main one: patients frequently must pay out of pocket to access care, incurring expenses that can be financially catastrophic. Poor referral systems that may not support timely pathways to care or adequate treatment, palliative or supportive services. Geographical inequities also exist in terms of Access to care (Brandão, 2021).

Tumor and Robotics

The field of medicine has also been invaded by robots. They weren't there to replace the doctors and Nurses but to help them in their routine work with them accurate tasks. Medical Robots Promise- J that really took off in the '90s. where it had a wide range of medical applications featured: laboratory robotics, telesurgery, surgery training, telesurgery, telemedicine and tele-consulting, rehabilitation, deaf assistance, and hospital robots. Medical robots assisting in surgeries for patients with heart attacks and made it possible to adjust the exact millimeters of prosthetic limbs. However, there are many challenges in the wide application of robotics in the medical field, mainly due to issues such as safety, accuracy, cost, and reluctance to accept this technique. Medical robots include a number of devices used in surgery and medical training, rehabilitation therapy, prosthetics and assistance (Campbell, 2019). The types of cancers for which we can use robotic surgery include gastrointestinal cancers, colorectal cancers, gynecological cancers, head and neck cancers, lung cancers, and urinary tract cancers (Campbell, 2019).

Tumor Diagnosis

The pathologist is forever trying to improve upon the accuracy of the diagnosis. In order that a method may fulfill the latter purpose it must be rapid, provide good staining differential qualities. Several imaging tests are used to conduct a diagnosis for brain tumors (Bai, 2020). These include:

- Neurological Examination
- Cerebral Angiogram
- MRI
- MRI Spectroscopy
- MRI Contrast
- Perfusion MRI
- Functional MRI

Types of Tumors: It exists different types of tumors such as:

Brain Tumor

A brain tumor is defined as an abnormally growing cluster of cells within the brain. Glioblastoma is a tumor in the malignant group that can develop in the brain and spinal cord, consisting of glia that supports nerve cells. Glioblastoma is mostly sporadic, except in rare cases like Turcot syndrome or Li-Fraumeni syndrome. It is one of the most dangerous and fatal diseases for humans, it must be detected, diagnosed, and cured in the early stages. It occurs in almost every part of the human body, but the human brain is the part where the tumor is found the most. Brain tumors can be considered dangerous since the majority of persons die in 9–12 months of identification.

Brain tumors occur in about 40,000–50,000 individuals each year, of which 20% are children (Guo, 2018). And according to Richters (2019) brain cancer is classified based on the level of cancerous growth present. There are two major types LGG (Low-Grade Glioma) and HGG (high-grade gliomas). Low grade gliomas are benign tumors (grade I or II). All low-grade gliomas eventually progress to high grade glioma and death. GBM(Glioblastoma) is the most common type of HGG, it is the most aggressive cancer that begins within the brain and is the most common type of malignant brain tumor among adults. It is usually aggressive, which means it can grow fast and spread quickly. Although there is no cure, there are treatments to help ease symptoms. The conventional method for medical resonance brain image classification and tumor detection is by human inspection. There is a need to have a fast and reliable method to classify if the tumor present is cancerous (LGG or HGG) (Lapointe, 2018). The common types of treatments used for a brain tumor are many and the treatment options and recommendations depend on several factors:

- The size, type, and grade of the tumor
- Whether the tumor is putting pressure on vital parts of the brain
- If the tumor has spread to other parts of the CNS or body
- Possible side effects
- The patient's preferences and overall health

Treatment options include those described below, such as surgery, radiation therapy, chemotherapy, and targeted therapy (Hotchkiss, 2021).

Surgery

Surgery is the removal of the tumor and some surrounding healthy tissue during an operation. It is usually the first treatment used for a brain tumor. It is often the only treatment needed for a low-grade brain tumor. Removing the tumor can improve neurological symptoms, provide tissue for diagnosis and genetic analysis, help make other brain tumor treatments more effective, and, in many instances, improve the prognosis of a person with a brain tumor.

Radiation Therapy

Radiation therapy is the use of high-energy x-rays or other particles to destroy tumor cells. Treatments using medication are used to destroy cancer cells. Medication may be given through the bloodstream to reach cancer cells throughout the body. When a drug is given this way, it is called systemic therapy. Medication may also be given locally, which is when the medication is applied directly to the cancer or kept in a single part of the body (Sahiner, 2019).

Lung Tumor

Lung cancer occurs when cells divide in the lungs uncontrollably. This causes tumors to grow. These can reduce a person's ability to breathe and spread to other parts of the body. Lung cancer is the third most common-Trusted Source cancer and the main cause Trusted Source of cancer-related death in the United States. It is most common in males, and in the U.S., Black males are around 15% more likely to develop it than white males. Smoking is a major risk factor, though not everyone who develops lung cancer has a history of smoking. Lung cancer can be fatal, but effective diagnoses and treatments are improving the outlook. The two main types of lung cancer are small cell lung cancer and non-small cell lung cancer, depending on how they appear under a microscope. Non-small cell lung cancer is more common than small cell lung cancer.

Treatment for Lung Cancer

In the past 10 years, there's been a lot of progress in lung cancer treatments. People are usually given more than one treatment at a time and you might have several courses of treatment. The outcome of lung cancer has gradually improved over recent years. Survival is closely related to stage of disease. If the cancer is detected and treated at an early stage, more people will survive for longer.The main treatments for lung cancer are:

1. Surgery There are a few different types of surgery. The surgeon might remove a section of your lung or your whole lung.
2. chemotherapy: This is medication that attacks cancer cells. They may be delivered straight into your bloodstream through a drip or you might have injections or tablets.
3. Targeted treatments: These are medicines, such as erlotinib, gefitinib and crizotinib, that stop the genetic mutations that cause some types of lung cancer. They can be very effective.

Figure 1. Anatomy of lung (Sundjaja, 2021)

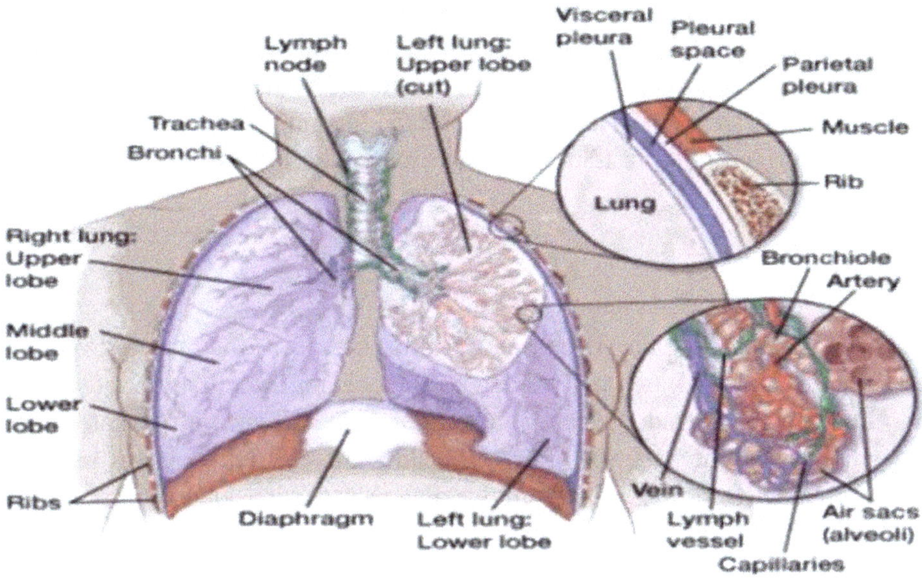

4. immunotherapies: These are medicines that work on the immune system in the body to enhance its response to cancer cells. An example is pembrolizumab, which can be used on its own or in combination with chemotherapy to treat metastatic NSCLC.
5. Radiotherapy: This treatment uses high-energy X-rays to destroy cancer cells. This can be both curative and palliative – helping to manage symptoms (Maly, 2019).

Prostate Cancer

The prostate is a walnut sized gland located between the bladder and external urinary sphincter through which the urethra passes. The primary function of the prostate is to produce a slightly alkaline fluid that protects spermatozoa from the acidic environment found within the vaginal tract (Rebello, 2021).

Prostate cancer accounts for 7% of newly diagnosed cancers in men globally. Annually, >1.2 million new cases are diagnosed and prostate cancer-related deaths exceed 350,000, making the disease one of the leading causes of cancer-associated death in men. In the USA, men of African or Caribbean descent have a twofold higher relative risk of early, more aggressive disease than white men. For some groups, such as Ashkenazi Jews and those of Icelandic descent, the risk of early, more aggressive disease is linked to germline mutations in genes such as BRCA2.

For others, the reasons for disparities in prostate cancer incidence and/or mortality are not known.

Prostate cancer is the second most common cancer in men. Genetic mutations in basal or luminal prostate epithelial cells are considered the primary driver of disease. Early diagnosis can result in favourable outcomes and high long-term survival but prognosis for men with castration-resistant metastatic disease is relatively poor (Rebello, 2021).

Diagnosis

Digital rectal examination (DRE), serum prostate-specific antigen (PSA) level measurement and MRI are standard diagnostic tools for detecting prostate cancer. Both DRE and PSA findings can be abnormal in the absence of prostate cancer or normal despite the presence of prostate cancer. A prostate biopsy is indicated for abnormal DRE results and a definitive diagnosis depends on histopathological verification. Prostate cancer aggressiveness is graded according to histological tumour features using the Gleason grade system, which was updated in 2014 to improve differential prognoses. The new Gleason grade group system, together with PSA levels and clinical tumour stage, is recommended for patient risk classification, owing to its superiority in predicting risk of potentially lethal disease.

Treatments for prostate cancer that has not spread elsewhere in the body are surgery or radiation therapy (RT), with or without hormone therapy. Active surveillance is also an option for men who have a low risk of their cancer spreading (Rebello, 2021).

Lymph Node

A small bean-shaped structure that is part of the body's immune system. Lymph nodes filter substances that travel through the lymphatic fluid, and they contain lymphocytes (white blood cells) that help the body fight infection and disease. There are hundreds of lymph nodes found throughout the body. They are connected to one another by lymph vessels.

Lymph nodes are part of the lymph system, a network of organs, nodes, ducts, and vessels that support the body's immune system. Nodes are little filters throughout the body. The cells in lymph nodes help to destroy infection, such as from a virus, or harmful cells, such as cancer cells (Richters, 2019). Cancer can start in the lymph nodes. This is called lymphoma. There are several types of lymphomas, such as non-Hodgkin lymphoma. Cancer cells can also spread to the lymph nodes from a cancer in any part of the body. This is called metastatic cancer. Cancer cells break off from a tumor in the body and travel to an area of lymph nodes. The cancer cells often travel to nodes near the tumor first (Schudel, 2019)

Figure 2. The complet system lymphatic (Schudel, 2019)

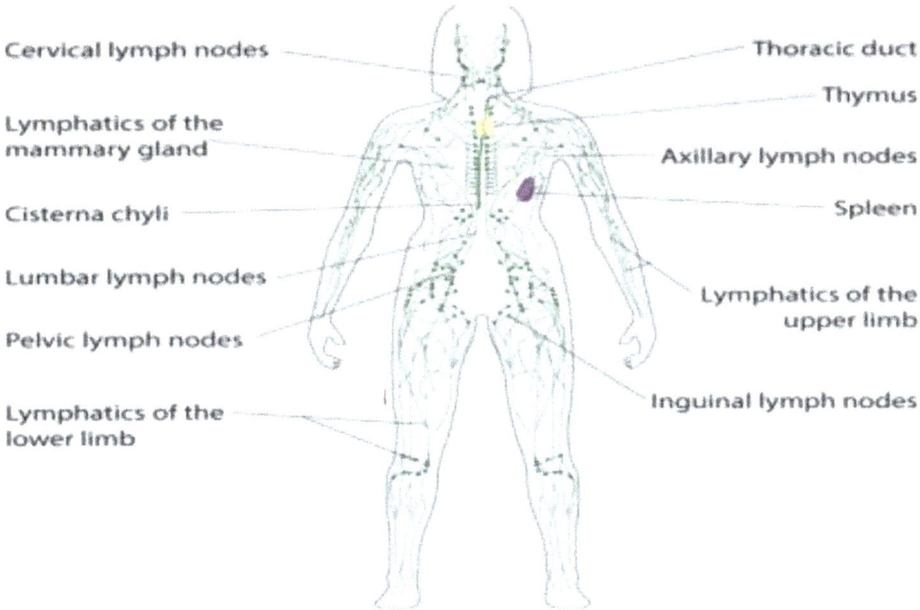

HISTOPATHOLOGY USING IN DEALING WITH CANCERS

Histopathology is widely used to study the manifestation of diseases. Histopathology is the branch of pathology that deals with microscopic examination of tissues. The thin tissue specimen collected from patients' body has to be converted to slides to make them feasible for microscopic examination. The above figure depicts the manual process of finding cancer through scanned images. This laborious process involves 5 stages:

- Tissue fixation: Tissues have the tendency to destroy and destruct themselves under the action of certain enzymes, to prevent such action it needs to be fixed. Collected tissue specimens are fixed using formalin. It terminates ongoing biochemical reactions providing mechanical strength and stability to the treated tissues.
- Specimen transfer to cassettes: The fixed tissue are then transferred to cassettes (special containers for tissue)
- Tissue processing: In this stage the tissue is prepared for sectioning (slicing). The tissue is first dehydrated and then cleared followed by infiltrating the tissue in an embedding agent like paraffin wax. Solidification paraffin wax

Figure 3. Image segemntation of histopathological tissue (Alifrangis, 2019)

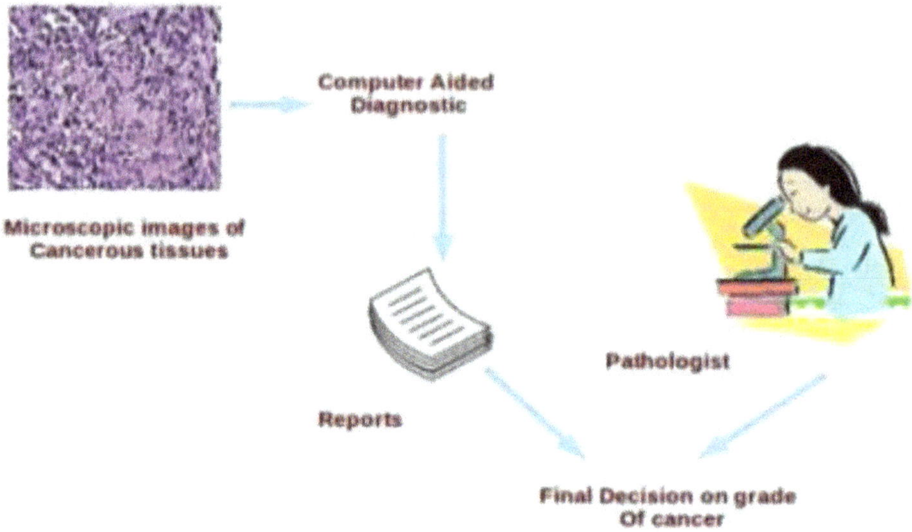

solidifies provides a support matrix for tissue specimen and now it can be sectioned.

- Sectioning: The tissue specimen can now be sectioned at different thickness and placed over a slide.
- Staining: Most cells are transparent, and appear almost colorless when unstained. Wide range of histochemical stains is therefore used to provide contrast to tissue sections, making tissue structures more visible and easier to evaluate (Alifrangis, 2019)

IMAGE SEGMENTATION

To segment an image means, to find its homogeneous regions and its contours, these two latter are assumed to be relevant, i.e. the regions must correspond to the parts significant real-world objects, and the contours at their apparent boundaries. Let “be the domain of the image and’ ’ the function which associates a value f(x,y) with each pixel (Minaee, 2021).

Figure 4. Conventional segmentation methods. Top row: edge-based method. Bottom row: region-based method

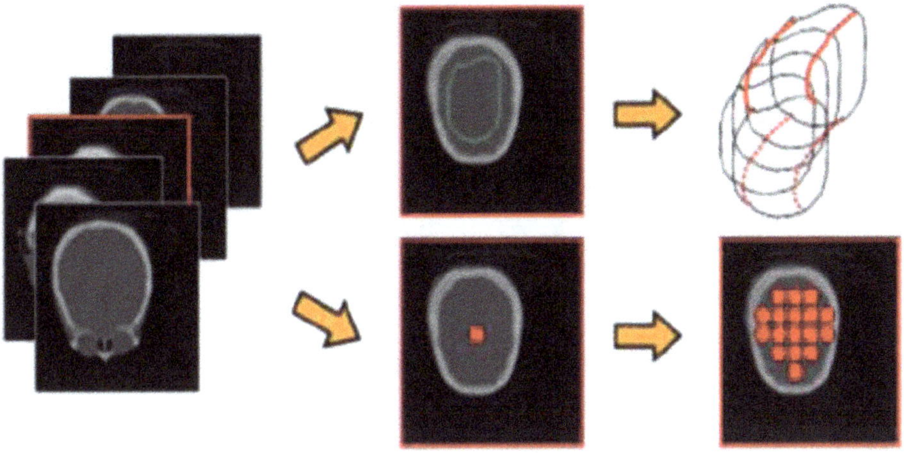

Edge-Based Segmentation

Segmentation can also be done by using edge detection techniques. In this technique, the frontier is identified to the segment. It is detected to identify the discontinuities in the image. Edges of the region are traced by identifying the pixel value and it is compared with the neighboring pixels. The edge-based segmentation there is no need for the detected edges to be closed. These methods have problems with images that are:

- Edge-less
- Very noisy
- Boundary that is very smooth
- Texture boundary (Priyadharsini, 2019)

Algorithm:

- Step1: Smooth the input image with a Gaussian filter.
- Step2: Compute the gradient and angle images.
- Step3: Apply maximum suppression to the gradient magnitude image.
- Step4: Use double thresholding and connectivity analysis to detect and link edges.

Figure 5. Segmentation based on regions (Ramesh, 2021)

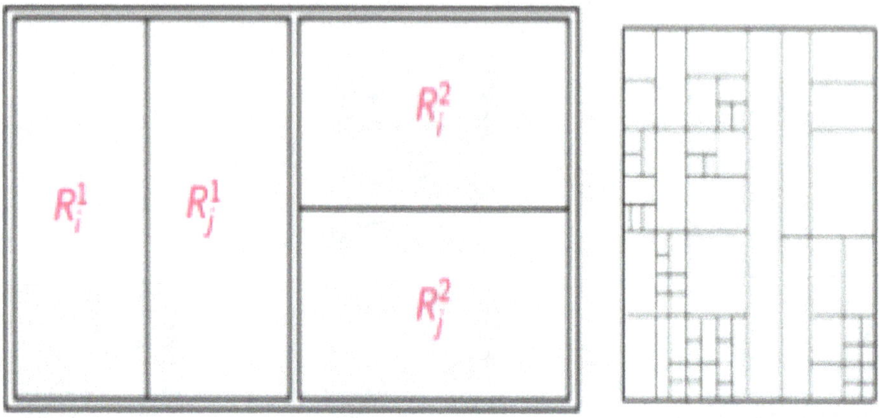

Advantage:

- The approach had to do with the way humans segment images.
- Works well in frames with good contrast between object and background.

Disadvantage:

- Not working correctly in images with smooth transitions and low contrast
- Sensitive to noise.
- Robust edge linking is not trivial.

Region-Based Segmentation

Locale developing may be a simple region-based picture division strategy. It is additionally classified as the pixel-based picture division method. The fundamental objective of division is to the division of the picture into districts. A few division strategies such as "Thresholding" accomplish this objective by looking at for the boundaries between locales based on discontinuities in gray levels or color properties (Ramesh, 2021).

Region growing Algorithm:

Step1: Starts with a set of seeds (starting pixels)

- ◦ Predefined seed
- ◦ All pixels as seeds

- Randomly chosen seeds

Step 2: Region growing intensity steps (bottom-up method)

- Find starting points
- Include nearest pixels with similar features (Gray level, texture, color)
- A similarity measure must be selected

Advantage:

- Region developing strategies can accurately partitioned the regions.
- Region developing strategies can give the first images.
- We can select the numerous criteria at the same time.

Disadvantage:

- The computation is solid, Dismissal matter the time or power.
- Excessive or variance escalation may result in excessive cleavage.
- This strategy may not recognize the shading of the genuine (Lou, 2020)

Watershed Segmentation

The presented the watershed change as a implies to isolating covering objects. A watershed is shaped by „flooding an picture from its nearby minima and shaping dams where waterfronts meet. When recreating this prepare for picture division, two approaches are utilized. One finds bowl at that point watersheds by taking a set of complement. One computes a total segment of the picture into bowls and subsequently refined the watersheds by boundary location. To be more express utilize the expression 'watershed transform' to represent a labeling of the picture. All focuses of a given catchment bowl have the same special name. An uncommon name, particular from all the names of the catchment bowls, is relegated to all focuses of the watersheds (Lou, 2020).

Threshold Based Segmentation

Thresholding is one of the only approaches for picture division based on concentrated levels. The Threshold-based procedure works on the explanation that the pixels falling inside the certain extend of escalated values. Speaks to one class and remaining pixels within the picture speaks to the another lesson. Thresholding can be actualized either locally or universally. For worldwide thresholding brightness edge esteem is to be chosen. To portion the picture into the protest and the background. It produces a parallel picture from given input picture. The pixels satisfying limit test are considered as protest pixels with twofold esteem „1 and another pixels are treated as background pixels with parallel esteem „0. Where T could be a predefined limit (Zebari, 2020).

Figure 6. An example of watershed segmentation: a) Original image; b) Gradient image; c) Watershed segmentation without using region markers (oversegmentation); d) Watershed segmentation using region markers (Lou, 2020)

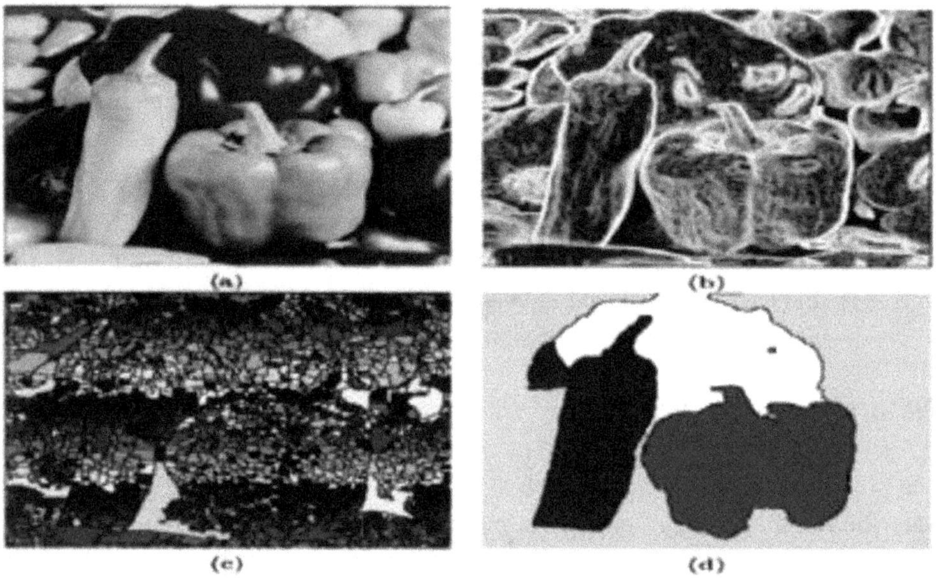

Clustering Based Color Image Segmentation

In cluster based division, information is combined into bunches. The information with comparative highlights will drop into one bunch. The information clusters are being distinctive from each other. The k-means calculation is commonly utilized for deciding the organization of the information. This unsupervised clustering approach features a solid partiality to induce caught into nearby minima. Creating an ideal arrangement. It makes clustering entirely dependent on the essential cluster centers dispersion. Think about to recognize adjust input parameters for getting ideal or problematic clustering results (Farshi, 2020).

CONCLUSION

In this chapter, we saw general concepts about bioinformatics, cancerous tumors, their types, diagnosis, and image segmentation and a quick overview of the different segmentation technologies in general. For the future of bioinformatics, we can found different Challenging issues that span all areas of modelling Systems

- Integrating data and developing models of complex systems across multiple spatial and temporal scales. Scale relations and coupling.
- Structure-function relationships.
- Image analysis and visualization.
- Basic mathematical issues.
- Data management.

REFERENCES

Alifrangis, C., McGovern, U., Freeman, A., Powles, T., & Linch, M. (2019). Molecular and histopathology directed therapy for advanced bladder cancer. *Nature Reviews. Urology*, *16*(8), 465–483. doi:10.103841585-019-0208-0 PMID:31289379

Anderson & Goodman. (2002). *Ethics and information technology: A case-based approach to a health care system in transition*. Springer.

Bai, X., Wang, Y., Song, Z., Feng, Y., Chen, Y., Zhang, D., & Feng, L. (2020). The basic properties of gold nanoparticles and their applications in tumor diagnosis and treatment. *International Journal of Molecular Sciences*, *21*(7), 2480. doi:10.3390/ijms21072480 PMID:32260051

Brandão, M., Guisseve, A., Bata, G., Firmino-Machado, J., Alberto, M., Ferro, J., Garcia, C., Zaqueu, C., Jamisse, A., Lorenzoni, C., Piccart-Gebhart, M., Leitão, D., Come, J., Soares, O., Gudo-Morais, A., Schmitt, F., Tulsidás, S., Carrilho, C., & Lunet, N. (2021). Survival impact and cost-effectiveness of a multidisciplinary tumor board for breast cancer in Mozambique, Sub-Saharan Africa. *The Oncologist*, *26*(6), e996–e1008. doi:10.1002/onco.13643 PMID:33325595

Campbell, R. G. (2019, December). Robotic surgery of the anterior skull base. *International Forum of Allergy & Rhinology*, *9*(12), 1508–1514. doi:10.1002/alr.22435 PMID:31633887

Farshi, T. R., Drake, J. H., & Özcan, E. (2020). A multimodal particle swarm optimization-based approach for image segmentation. *Expert Systems with Applications*, *149*, 113233.

Guo, R., Zhao, Y., Zou, Q., Fang, X., & Peng, S. (2018). Bioinformatics applications on apache spark. *GigaScience*, *7*(8), giy098. PMID:30101283

Hotchkiss, K. M., & Sampson, J. H. (2021). Temozolomide treatment outcomes and immunotherapy efficacy in brain tumor. *Journal of Neuro-Oncology*, *151*(1), 55–62. doi:10.100711060-020-03598-2 PMID:32813186

Lapointe, S., Perry, A., & Butowski, N. A. (2018). Primary brain tumours in adults. *Lancet*, *392*(10145), 432–446. doi:10.1016/S0140-6736(18)30990-5 PMID:30060998

Lou, S., Pagani, L., Zeng, W., Jiang, X., & Scott, P. J. (2020). Watershed segmentation of topographical features on freeform surfaces and its application to additively manufactured surfaces. *Precision Engineering*, *63*, 177–186. doi:10.1016/j.precisioneng.2020.02.005

Malik, H., Fatema, N., & Alzubi, J. A. (Eds.). (2021). *AI and machine learning paradigms for health monitoring system: intelligent data analytics*. Springer. doi:10.1007/978-981-33-4412-9

Maly, V., Maly, O., Kolostova, K., & Bobek, V. (2019). Circulating tumor cells in diagnosis and treatment of lung cancer. *In Vivo, 33*(4), 1027-1037.

Minaee, S., Boykov, Y. Y., Porikli, F., Plaza, A. J., Kehtarnavaz, N., & Terzopoulos, D. (2021). Image segmentation using deep learning: A survey. *IEEE Transactions on Pattern Analysis and Machine Intelligence*, 1. doi:10.1109/TPAMI.2021.3059968 PMID:33596172

Priyadharsini, R., & Sharmila, T. S. (2019). Object detection in underwater acoustic images using edge based segmentation method. *Procedia Computer Science*, *165*, 759–765. doi:10.1016/j.procs.2020.01.015

Ramesh, K. K. D., Kumar, G. K., Swapna, K., Datta, D., & Rajest, S. S. (2021). A review of medical image segmentation algorithms. *EAI Endorsed Transactions on Pervasive Health and Technology*, *7*(27), e6–e6.

Rebello, R. J., Oing, C., Knudsen, K. E., Loeb, S., Johnson, D. C., Reiter, R. E., Gillessen, S., Van der Kwast, T., & Bristow, R. G. (2021). Prostate cancer. *Nature Reviews. Disease Primers*, *7*(1), 1–27. doi:10.103841572-020-00243-0 PMID:33542230

Richters, M. M., Xia, H., Campbell, K. M., Gillanders, W. E., Griffith, O. L., & Griffith, M. (2019). Best practices for bioinformatic characterization of neoantigens for clinical utility. *Genome Medicine*, *11*(1), 1–21. doi:10.118613073-019-0666-2 PMID:31462330

Sahiner, B., Pezeshk, A., Hadjiiski, L. M., Wang, X., Drukker, K., Cha, K. H., Summers, R. M., & Giger, M. L. (2019). Deep learning in medical imaging and radiation therapy. *Medical Physics*, *46*(1), e1–e36. doi:10.1002/mp.13264 PMID:30367497

Schudel, A., Francis, D. M., & Thomas, S. N. (2019). Material design for lymph node drug delivery. *Nature Reviews. Materials*, *4*(6), 415–428.

Sundjaja, J. H., & Bordoni, B. (2021). Anatomy, Thorax, Lung Veins. StatPearls Publishing.

Whitman, W. B., Chuvochina, M., Hedlund, B. P., Hugenholtz, P., Konstantinidis, K. T., Murray, A., Palmer, M., Parks, D. H., Probst, A. J., Reysenbach, A.-L., Rodriguez-R, L. M., Rossello-Mora, R., Sutcliffe, I., & Venter, S. N. (2022). Development of the SeqCode: A proposed nomenclatural code for uncultivated prokaryotes with DNA sequences as type. *Systematic and Applied Microbiology*, *45*(5), 126305. doi:10.1016/j.syapm.2022.126305 PMID:36049255

Zebari, D. A., Zeebaree, D. Q., Abdulazeez, A. M., Haron, H., & Hamed, H. N. A. (2020). Improved threshold based and trainable fully automated segmentation for breast cancer boundary and pectoral muscle in mammogram images. *IEEE Access: Practical Innovations, Open Solutions*, *8*, 203097–203116. doi:10.1109/ACCESS.2020.3036072

Chapter 11
Computer Vision in Healthcare

Cheimaa Bentadj
Dr. Moulay Tahar University, Algeria

ABSTRACT

Computer vision has made substantial progress during the past several decades, and it has covered many areas including healthcare. This chapter introduces deep learning techniques for healthcare with an emphasis on the discussion of deep learning in computer vision, the authors describe how these techniques can impact a few key areas of health precaution that include wearing a mask and respecting social distancing to reduce the spread of airborne epidemics. This chapter provides an overview of deep learning techniques that can detect social distancing and mask wearing through modern surveillance technologies, including drones and surveillance cameras.

INTRODUCTION

Artificial intelligence is a broad branch of computer science, the goal of IA is to create systems that can function intelligently and independently, the purpose of artificial intelligence is to make the machine learn by itself without human guidance and to think like humans, and to mimic human consciousness. The ideal feature of IA is its ability to make a decision that has the best chance of achieving a specific goal, techniques from the field of artificial intelligence, and more specifically deep learning methods, have been core components of most recent developments in the field of computer vision. Where it is already being exploited to solve the problems and diseases of the times like COVID-19.

The COVID-19 pandemic poses a number of challenges to the Artificial Intelligence (AI) Community, among these challenges are use it for social control

DOI: 10.4018/978-1-6684-5656-9.ch011

and detection of face mask violation and social distance, use it in the search for treatments and a vaccine, also to detect the violation of protective equipment, all of these using a Convolutional Neural Network (CNN).

COVID-19 PANDEMIC DEFINED

"We now have a name for the disease caused by coronavirus and it's COVID-19", said Dr Tedros Adhanom Ghebreyesus, Director-General of WHO on Feb 11, 2020 (BBC News, 2020), World Health Organization (WHO) recently updated the name novel coronavirus pneumonia, previously named by Chinese scientists (Wang, 2020), to coronavirus disease 2019 (COVID-19).

More attention should be paid to comorbidities in the treatment of COVID-19. In the literature, COVID-19 is characterized by the symptoms of viral pneumonia such as fever, fatigue, dry cough, and lymphopenia. Many of the older patients who become severely ill have evidence of underlying illness such as cardiovascular disease, liver disease, kidney disease, or malignant tumours (Wang, 2020).

Humanity is currently facing a new type of Coronavirus, where the first patients with coronavirus disease 2019 (COVID-19) were recorded in December 2019 in China (de Groot, 2020), since then, the Coronavirus has spread globally.

COVID-19 AND ITS IMPACT ON SOCIETY

Social distancing involves staying away from people to avoid the spreading and catching the virus. It is a new emerging terminology which means to avoid the crowd. This has forced people to work from home and avoid social gatherings and contacting even their near ones (Singh, 2020).

Eric Kleinberg, a New York University sociologist, stated that "we've also entered a new period of social pain. There's going to be a level of social suffering related to isolation and the cost of social distancing that very few people are discussing this yet."

Statistics WHO indicate that at least 165,000 have succumbed to the disease, most patients with COVID-19 suffer from severe respiratory problems, and elderly people in particular or persons with comorbidities seem to be at greatest risk (Wang T, 2020).

In the following figure, we note the extent of the Coronavirus spread around the world

Figure 1. Countries with confirmed Corona Virus Cases

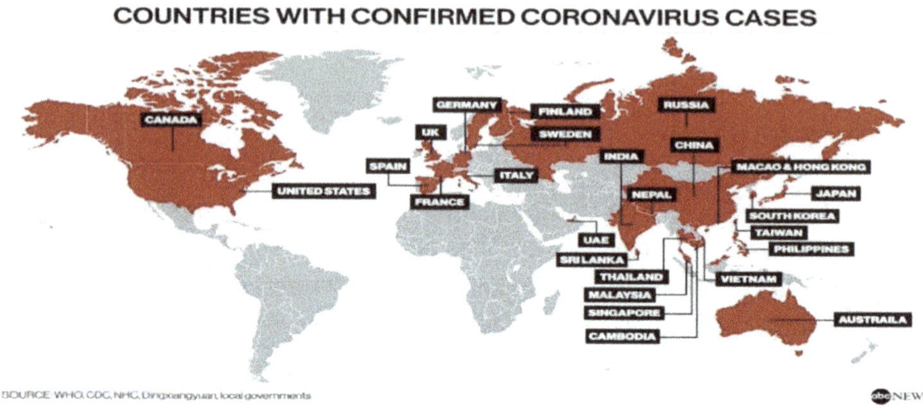

CORONAVIRUS PREVENTION

Prevention is better than cure, so it is best to follow the guidelines required by the World Health Organization to stay safe by taking some simple precautions, such as physical distancing, wearing a mask, keeping rooms well ventilated, avoiding crowds, cleaning your hands, and coughing into a bent elbow or tissue. Check local advice where you live and work do it all!

Maintain at least a 1-meter distance between yourself and others to reduce your risk of infection when they cough, sneeze or speak. Maintain an even greater distance between yourself and others when indoors. The further away, the better.

Make wearing a mask a normal part of being around other people. The appropriate use, storage and cleaning or disposal are essential to make masks as effective as possible. And need for sterilization as well.

MACHINE LEARNING (ML)

ML is a subset of artificial intelligence, this means that computer programs can learn automatically from new data and adapt to it, without consulting or being assisted by humans.

It's building a mathematical model based on sample data, known as "training data," in order to make predictions or decisions without being explicitly programmed to perform the task (Zhang, 2020).

Machine learning can be divided into several categories, can be supervised learning, unsupervised learning, reinforcement learning.

Figure 2. The Underfitting and Overfitting

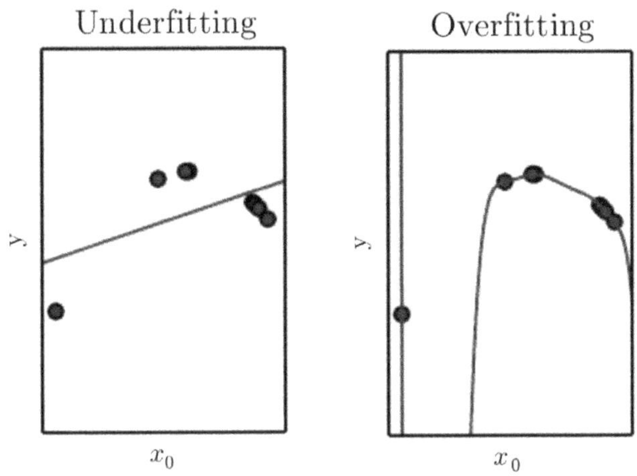

Overfitting (OF), and Underfitting (UF)

These two factors correspond to the two central challenges in machine learning: underfitting and overfitting. Underfitting occurs when the model is not able to obtain a sufficiently low error value on the training set. Overfitting occurs when the gap between the training error and test error is too large (Goodfellow, 2016).

The equation of underfitting is simple and not able to cover all points. It has a big problem in predicting the upcoming values and the loss function is large.

The mathematical formula for overfitting used adjusts all points, which causes a problem in prediction due to the unhelpful excess of data.

Supervised Learning

It is the process of training a model by entering data as an input, as well as correct output data, the term of "labeled data" is usually referred to the input/output pair.

The learning algorithm is given labeled data and the computer builds relationships and patterns between the data to predict the outputs when we give it a new input.

Supervised learning is often used to create machine learning models for two types of problems, regression and classification, the first, its outputs are real numbers and as for the second, its outputs are classes.

Unsupervised Learning

In unsupervised learning, unlabeled data are used to predict unknown results and the algorithm has to determine how to discover the hidden patterns in the dataset (Krittanawong, 2017), without the need for human intervention, so the main task is to find the solutions on its own.

Unsupervised learning models are utilized for two main tasks, clustering and dimensionality reduction.

Reinforcement Learning

It's a learning paradigm concerned with learning to control a system so as to maximize a numerical performance measure that expresses a long-term objective (Szepesvári,2010), Safe Reinforcement Learning can be defined as the process of learning policies that maximize the expectation of the return in problems in which it is important to ensure reasonable system performance and/or respect safety constraints during the learning and/or deployment processes (Garcıa,2015).

Deep learning (DL)

DL is a form of machine learning that enables computers to learn from experience and understand the world in terms of a hierarchy of concepts. Because the computer gathers knowledge from experience (Goodfellow, 2016), it is a particular case of machine learning but at a deep level, If we make a simple comparison between deep learning and machine learning, we will find that the more data, the Performance increased in deep learning, as for machine learning, the more data, the more stable the performance. So in these chapters, we will focus on the deep learning field especially the Convolutional Neural Network.

Computer Vision (CV)

Computer vision (CV) is a study of how to use computer simulation of human visual science, its main task is through the collection of images (or video) analysis and understanding (Al-Saffar,2017), we highlight computer vision combined with artificial intelligence algorithms that achieved important results in image recognition (such as face recognition, handwritten characters), image classification, it is one of the areas that's been advancing rapidly thanks to deep learning. Deep learning computer vision is now helping self-driving cars figure out where the other cars and pedestrians are around so as to avoid them, at the same time, high-tech such as the intelligent robot and many other disciplines.

State of Computer Vision

Data Vs. Hand engineering

Figure 3. Data Vs. Hand engineering

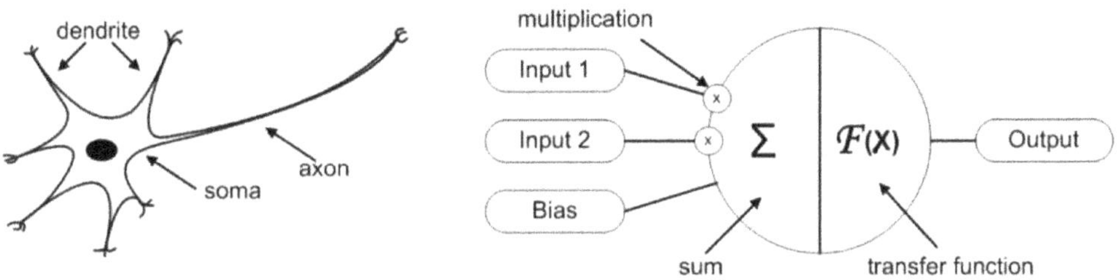

ARTIFICIAL NEURON (AN)

Artificial Neuron is a basic building block of every artificial neural network. Its design and functionalities are derived from observation of a biological neuron that is basic building block of biological neural networks (systems) which includes the brain, spinal cord and peripheral ganglia. Similarities in design and functionalities can be seen in Figure 4 (Krenker, 2011).

Where the left side of a figure represents a biological neuron with its soma, dendrites and axon and where the right side of a figure represents an artificial neuron with its inputs, weights, transfer function, bias and outputs (Krenker, 2011).

ARTIFICIAL NEURAL NETWORKS (ANNS)

Artificial Neural Networks (ANNs) are computational processing systems of which are heavily inspired by way biological nervous systems (such as the human brain) operate (Kalchbrenner, 2014), that mimic human operations and the way the brain analyzes and processes information Using the idea that brain cells think and analyze data, to design a similar algorithm, the idea has nothing to do with human thinking, but rather with the method of analysis.

Artificial neurons are designed to gather neurons in the brain, which contain multiple layers, the first having x1, ..., xn inputs, and the following layers consider the outputs of the previous layers an inputs to it, until the output layers that have a single neuron output or multi neuron output. The neuron output signal "a" is given by the following relationship:

Figure 4. Biological and artificial neuron design (Krenker, 2011)

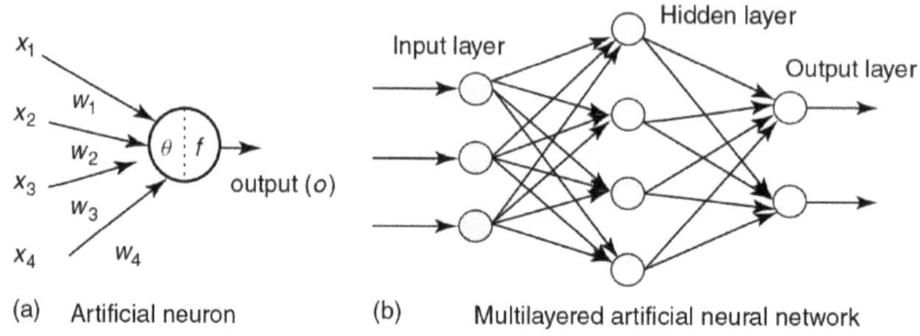

$$a = f(g) = f\left(\sum_{j=1}^{n} w_j x_j\right)$$

Where

w_j The weight vector

x_j The input vectors, and the function is $f(g)$ is referred to as an activation function

A typical artificial neuron and the modeling of a multilayered neural network are illustrated in Figure 5 (Abraham, 2005), In Figure "a" we have one layer with four inputs and one output, which is the activation function, in Figure "b" we have a simple three layer, comprised of an input layer, a hidden layer, and an output layer. This structure is the basis of a number of common ANN architectures,

Activation Function

Different activation functions can exist to activate the neuron they are listed in Table 1. The most used functions are: « Sigmoid », « Tanh », « ReLu », and «Softmax».

CONVOLUTIONAL NEURAL NETWORK (CNN)

Convolutional neural networks are deep learning algorithms that take input images and convolve them with filters or kernels to extract features. An NxN image is convolved with an fxf filter and this convolution operation learns the same feature on the entire image (Zeiler, 2014).

In recent years, the convolutional neural network (CNN) has emerged as the prevalent model for machine learning and computer vision. Now, deep CNNs are

Figure 5. Architecture of an artificial neuron and a multilayered neural network (Abraham, 2005)

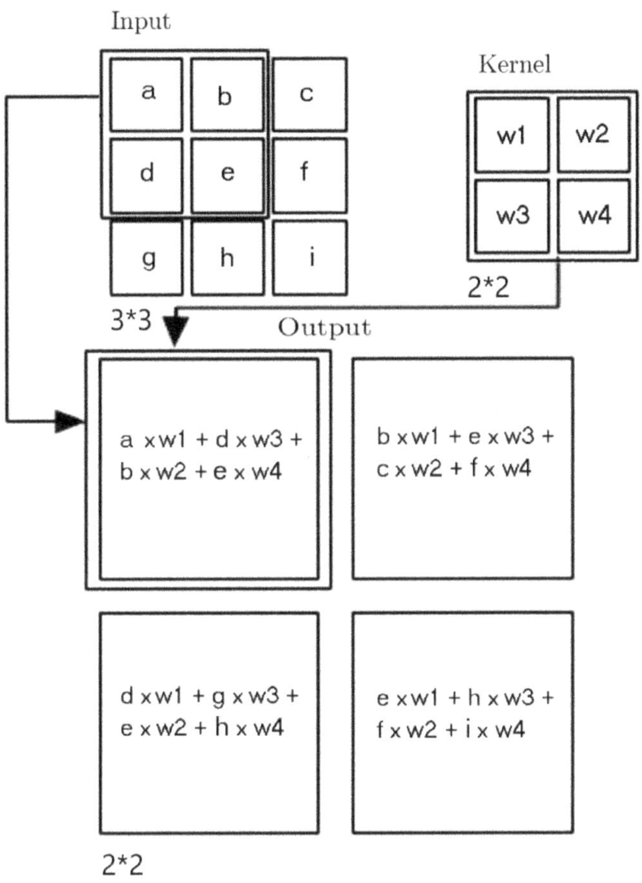

widely used in a broad range of real-life applications, such as image classification, object detection, and image segmentation (Li, 2019).

In the following figure, we have an example of 2-D convolution with a kernel.

Edge Detection

Edge detection is a fundamental image processing technique that involves computing an image gradient to quantify the magnitude and direction of edges in an image, image gradients are used in various downstream tasks in computer vision such as line detection, feature detection, and image classification (Mani, 2019).

More edge detection:

Table 1. Different activation functions

The name of function	Input / output relationship
Sigmoid	$\frac{1}{1+e^{-x}}$
Tanh	$Tanh(x) = \frac{1-e^{(-2x)}}{1+e^{(-2x)}}$
ReLU	$ReLu(x) = \max(x,0).$ $f(x) = x \; ,x \geq 0$ $f(x) = 0 \; ,x < 0$
SoftMax	$\sigma(z)_j = \frac{e^{zj}}{\sum_{k=1}^{k} e^{zk}}$ For j=1,…,K (Sharma,2017)

In the following figure, we represents more edge detection
In the following figure, we apply the edge detection using Sobel filters

Figure 6. an example of 2-D convolution with a kernel

Figure 7. different edge detection

1	1	1
0	0	0
-1	-1	-1

Horizontal edge detection

1	0	-1
2	0	-2
1	0	-1

Sobel filter (puts a little bit more weight to the central row, and this makes it maybe a little bit more robust) them

3	0	-3
10	0	-10
3	0	-3

Horizontal

Scharr filter

the both of them have an horizontal version

Figure 8. Edge detection using Sobel filters (Mani, 2019)

0	0	0	0	0	0	0	0
0	3	3	4	4	7	0	0
0	9	7	6	5	8	2	0
0	6	5	5	6	9	2	0
0	7	1	3	2	7	8	0
0	0	3	7	1	8	3	0
0	4	0	4	3	2	2	0
0	0	0	0	0	0	0	0

$6 \times 6 \rightarrow 8 \times 8$

*

1	0	-1
1	0	-1
1	0	-1

3×3

=

-10	2	1	-5	7	15
-15	3				

6×6

Padding

One of the drawbacks of the convolution step is the loss of information that might exist on the border of the image (Albawi, 2017), that the filter does not pass to all the pixels evenly, so a simple and effective way to solve the problem is to use a zero padding, and this makes it pass by all the borders at least three times, while it was only once, and also it keeps the dimensions of the input matrix.

For example, in Figure 9, the input image was N = 6, after the padding N= 8, F = 3 and stride= 1, the result will be 6 x 6.

The Valid convolutions: "No padding" so ***NxN* * *FxF*** $(\boldsymbol{N}-\boldsymbol{F}+1)*(\boldsymbol{N}-\boldsymbol{F}+1)$

Figure 9. ConvNet application using a single padding

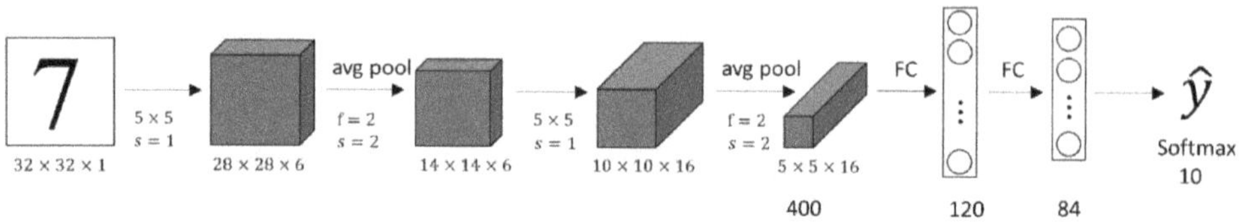

The Same convolutions: dimension of input matrix = dimension of output matrix

$$\left(\boldsymbol{N}+2\boldsymbol{P}-\boldsymbol{F}+1\right)*\left(\boldsymbol{N}+2\boldsymbol{P}-\boldsymbol{F}+1\right)=\boldsymbol{N}*\boldsymbol{N}$$

Through this equation we find this rule to achieve the same convolutions:

$$\boldsymbol{P}=\left(\boldsymbol{F}-1\right)/\,2$$

Strided convolutions: It's helpful for save time but the quality decreases

$$\left(\boldsymbol{NxN}\right)*\left(\boldsymbol{FxF}\right) \qquad \left(\left(\boldsymbol{N}+2\boldsymbol{P}-\boldsymbol{F}\right)/\,\boldsymbol{S}\right)+1*\left(\left(\boldsymbol{N}+2\boldsymbol{P}-\boldsymbol{F}\right)/\,\boldsymbol{S}\right)+1$$

Pooling

Max pooling: "Get the max value"

Averadge pooling: "Get the average value"

Classic model

LeNet-5:

LeNet was introduced by Yan LeCun for digit recognition. It includes five convolutional layers (Albawi, 2017), it's a simple and old network that contains only black and white images, as for the input image, the size is 32×32×1. The Typical architecture for LeNet-5 is presented in Figure 10.

LeNet-5 CNN architecture is made up of seven layers. The layer composition consists of 3 convolutional layers, two subsampling layers which are average pooling, and two fully connected layers.

AlexNet:

Table 1.

1	3	2	3
2	9	1	1
1	3	2	3
5	6	1	2
9	3		
6	2		

4*4
2*2

Table 2.

2	3	6	4
2	9	1	1
1	8	2	3
5	6	1	2
4	3		
5	2		

4*4
2*2

Figure 10. AlexNet architecture

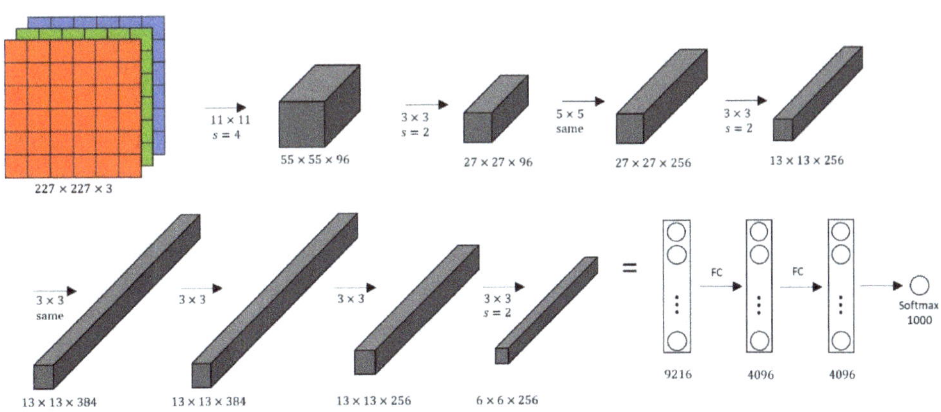

Figure 11. AlexNet architecture

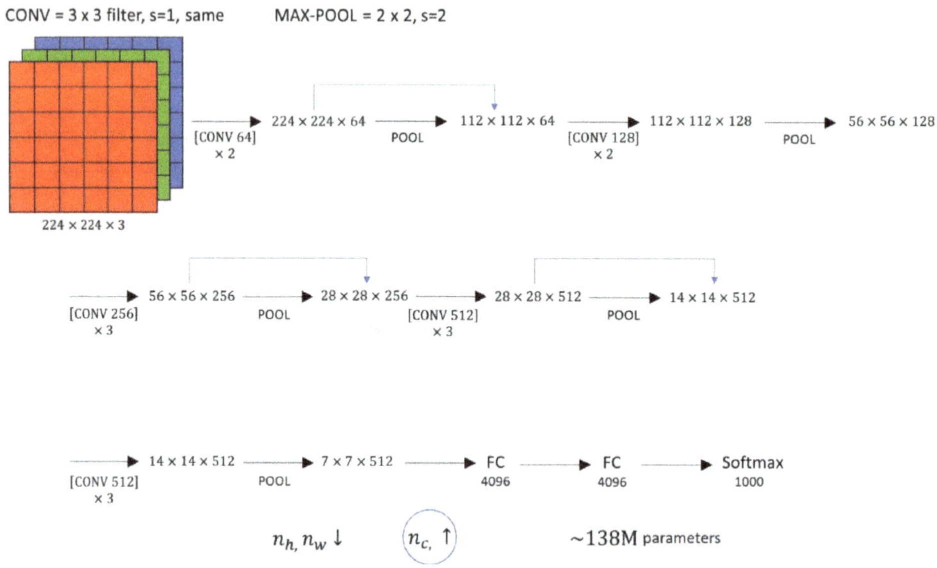

The AlexNet model is a Convolutional Neural Network consisting of eight learned layers, five convolutional layers, and three fully connected layers. The first layer is used to filter the input image, its size should be 227 x 227 x 3. The architecture of AlexNet is presented in Figure 11

VGG-16:

VGG16 architecture consists of 16 layers. It is considered to be one of the excellent vision model architecture till now. The most unique thing about VGG16 is that instead of having a large number of hyper-parameter they focused on having convolution layers of 3x3 filter with a stride 1 and always used the same padding and max-pooling layer of 2x2 filter of stride 2.

The remarkable thing about the VGG–16 is that instead of having so many hyperparameters we will use a much simpler network. We will focus on just having convNet layers that are just 3×3 filters with a stride of 1, and with the same padding. In all Carpooling layers, we will use 2×2 filters with a stride of 2. Number 16 in the name VGG–16 refers to the fact that this has 16 layers that have some weights. This is a pretty large network and has a total of about 138M parameters and we notice that the length and the width decrease in half, while the deep increases to double.

Figure 12. VGG-16 architecture

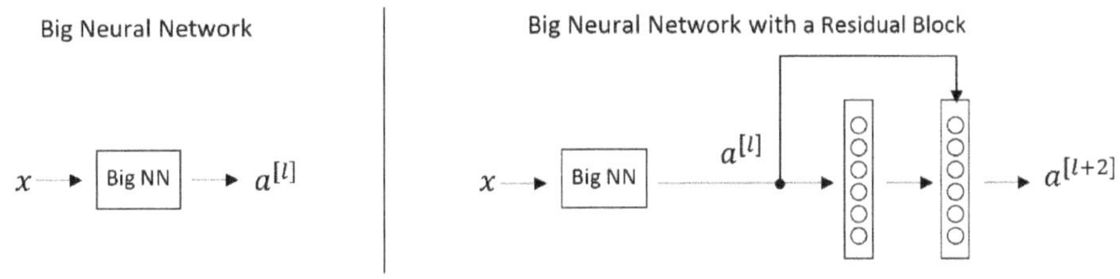

VGG−19 neural network is bigger than VGG−16, but because VGG−16 does almost as well as the VGG−19 a lot of people will use VGG−16. The architecture of VGG-16 is presented in Figure 12.

Residual Networks (ResNets)

The illustration Figure 13 represents how can ResNets work so well, Let's go through our example that explains how we can make the neural network deeper and deeper without really hurting our ability to get them to do well on the training set. Hopefully, doing well on the training set is usually a prerequisite to doing well on the test set. So, being able to at least train ResNet to do well on the training set is a good first step towards that.

- How does the ResNets work:

Figure 13. big NN with ResNets architecture

3	2	1	0	5	7
0	5	2	1	8	4
1	5	9	3	6	2
2	0	1	5	4	8
4	3	2	1	7	5
3	7	8	4	9	1

6*6

* 2

1*1

⟹

6	4	2	0	10	14
0	10	4	2	16	8
2	10	18	6	12	4
4	0	2	10	8	16
8	6	2	2	14	10
6	14	16	8	18	2

6*6

$a[l]$--------> Linear --------> ReLu --------> Linear --------> ReLu -------->

$a[l+1]$ $a[l+2]$

Short cut / Skip connection

$ReLu\ \ a>0$

$$\boldsymbol{a}[\boldsymbol{l}+2]=\boldsymbol{g}(\boldsymbol{z}[\boldsymbol{l}+2]+\boldsymbol{a}[\boldsymbol{l}])$$

$$=\boldsymbol{g}(\boldsymbol{b}[\boldsymbol{l}+2].\boldsymbol{a}[\boldsymbol{l}+1]+\boldsymbol{b}[\boldsymbol{l}+1]+\boldsymbol{a}[\boldsymbol{l}])$$

When we do the regularization for the values of $\boldsymbol{z}[\boldsymbol{l}+2]$ this leads to reduces the value $\boldsymbol{w}[\boldsymbol{l}+2]$ and that's mean increased the values of $\boldsymbol{a}[\boldsymbol{l}+1]$, Thus reducing the value $\boldsymbol{b}[\boldsymbol{l}+2]$

As the $\boldsymbol{w}[\boldsymbol{l}+1]$ approaches zero, the $\boldsymbol{a}[\boldsymbol{l}+1]$ increases dramatically, which makes the $\boldsymbol{b}[\boldsymbol{l}+2]$ also decrease to zero, which makes each of **z** devolve to zero, and here it becomes $\boldsymbol{a}[\boldsymbol{l}+2]=\boldsymbol{g}(\boldsymbol{a}[1])$

And because the ReLu function is equal to the value of the input, if it were positive, then that means increasing two layers using the ResNets, doesn't almost change the output Trainer error.

Networks in Networks and 1x1 Convolutions

We have an example of 1×1 convolution

It's seems like the same matrix with the dimension but each number multiply by 2, so it's useless in this case, 1x1 Convolutions be useful when we use it with a stereoscopic matrix.

Using 1 x 1 Convolution

In this example, we find that the use of the 1x1 convolutional method led to the number of parameters from 120M to 12,4M parameters, it is useful to reduce the number of parameters, in the following figure 12 we see how it works.

Figure 14. the role of 1 x 1 convolution in reducing parameters

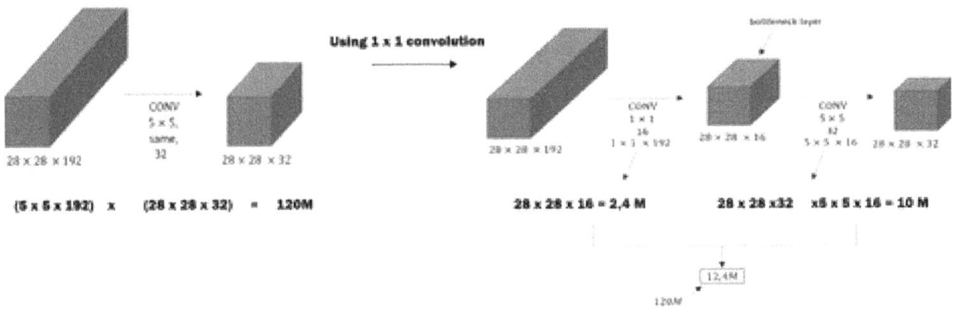

Transfer Learning

Many computer vision researchers have trained their algorithms on these datasets. Sometimes this training takes several weeks and might take many GPUs. The fact that someone else has done this task and gone through the painful high-performance research process means that we can often download open source weights. Next, we can use them as a very good parameter's initialization for our own neural network. That is, we can use transfer learning to transfer knowledge from some of these very large public datasets to our own problem with a freezing method.

Figure 15. illustration of how transfer learning works

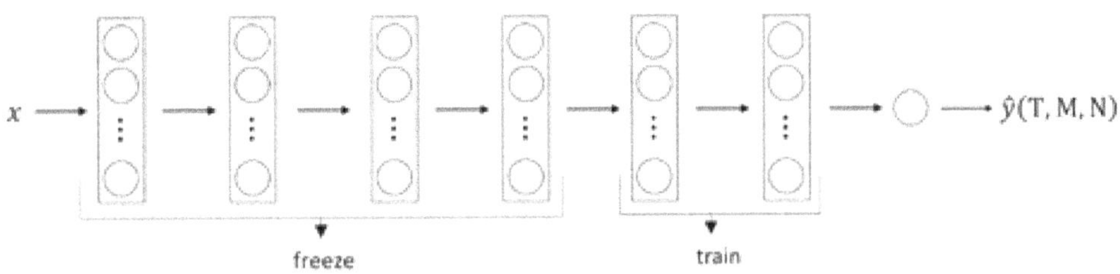

So in this Figure 13 we will freeze all the weights of the previous layers by calculating the activation function and saving them in our disk, sometimes we train the softmax layer only with our own issues with the new output, and sometimes we can change the last layers using SoftMax. Transfer learning has several benefits, but

Figure 16. Different augmentation methods

Type of data augmentation	Example
Mirroring	
Random cropping	
Color shifting	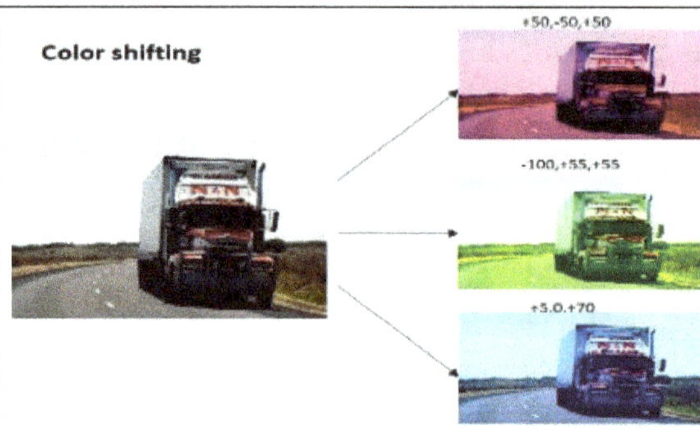

the main advantages are saving training time, better performance of neural networks (in most cases), and not needing a lot of data.

Data Augmentation

Most computer vision tasks could use more data and data augmentation is one of the techniques that is often used to improve the performance of computer vision systems and augment the accuracy of any algorithm. We have many data augmentation methods in computer vision:

We have also (Rotation, Shearing, local wraping)

Figure 17. a simple Dataset image

Dataset

The best way to make a machine learning model generalize better is to train it on more data (Goodfellow, 2016), it's better to create your own datasets than to use the data augmentation

- The MNIST Dataset:

The MNIST dataset consists of scans of handwritten digits and associated labels describing which digit 0–9 is contained in each image (Goodfellow, 2016), the dataset has 70,000 images to train and test the model. The training and test set

Figure 18. Fashion-MNIST Dataset

distribution is 60,000 train images and 10,000 test images. The size of each image is 28x28 pixels (784 pixels) (Chauhan, 2018).

- Fashion-MNIST Dataset:

The Fashion-MNIST dataset which consists of 60,000 training set pictures and 10,000 test set pictures. Each symbol is a gray-scale image of 28×28, linked to 10-category labels as shown in Figure 16 (Kayed, 2020)

CONCLUSION

In this chapter, we presented what is artificial intelligence, specifically deep learning, as we have highlighted the convolutional neural network and its characteristics.

REFERENCES

Abraham, A. (2005). Artificial neural networks. Handbook of measuring system design, 15.

Al-Saffar, A. A. M., Tao, H., & Talab, M. A. (2017, October). Review of deep convolution neural network in image classification. In *2017 International Conference on Radar, Antenna, Microwave, Electronics, and Telecommunications (ICRAMET)* (pp. 26-31). IEEE. 10.1109/ICRAMET.2017.8253139

Albawi, S., Mohammed, T. A., & Al-Zawi, S. (2017, August). Understanding of a convolutional neural network. In *2017 International Conference on Engineering and Technology (ICET)* (pp. 1-6). IEEE.

BBC News. (2020). *Coronavirus disease named Covid-19.* https://www.bbc.co.uk/news/world-asia-china-51466362

Chauhan, R., Ghanshala, K. K., & Joshi, R. C. (2018, December). Convolutional neural network (CNN) for image detection and recognition. In *2018 First International Conference on Secure Cyber Computing and Communication (ICSCCC)* (pp. 278-282). IEEE. 10.1109/ICSCCC.2018.8703316

de Groot, N. G., & Bontrop, R. E. (2020). *COVID-19 pandemic: is a gender-defined dosage effect responsible for the high mortality rate among males?* Academic Press.

Garcıa, J., & Fernández, F. (2015). A comprehensive survey on safe reinforcement learning. *Journal of Machine Learning Research, 16*(1), 1437–1480.

Goodfellow, I., Bengio, Y., Courville, A., & Bengio, Y. (2016). Deep learning: Vol. 1. *No. 2*. MIT Press.

Kalchbrenner, N., Grefenstette, E., & Blunsom, P. (2014). *A convolutional neural network for modelling sentences.* arXiv preprint arXiv:1404.2188.

Kayed, M., Anter, A., & Mohamed, H. (2020, February). Classification of garments from fashion mnist dataset using cnn lenet-5 architecture. In *2020 International Conference on Innovative Trends in Communication and Computer Engineering (ITCE)* (pp. 238-243). IEEE. 10.1109/ITCE48509.2020.9047776

Krenker, A., Bešter, J., & Kos, A. (2011). Introduction to the artificial neural networks. In *Artificial Neural Networks: Methodological Advances and Biomedical Applications*. InTech.

Krittanawong, C., Zhang, H., Wang, Z., Aydar, M., & Kitai, T. (2017). Artificial intelligence in precision cardiovascular medicine. *Journal of the American College of Cardiology, 69*(21), 2657–2664. doi:10.1016/j.jacc.2017.03.571 PMID:28545640

Li, Y., & Du, Y. (2019). A novel software-defined convolutional neural networks accelerator. *IEEE Access: Practical Innovations, Open Solutions, 7*, 177922–177931. doi:10.1109/ACCESS.2019.2956841

Mani, S. (2019). *Artistic Enhancement and Style Transfer of Image Edges using Directional Pseudo-coloring.* arXiv preprint arXiv:1906.07981.

Sharma, S. (2017). Activation functions in neural networks. *Towards Data Science, 6*.

Singh, J., & Singh, J. (2020). COVID-19 and its impact on society. *Electronic Research Journal of Social Sciences and Humanities, 2*.

Szepesvári, C. (2010). Algorithms for reinforcement learning. *Synthesis Lectures on Artificial Intelligence and Machine Learning, 4*(1), 1-103.

Wang, T., Du, Z., Zhu, F., Cao, Z., An, Y., Gao, Y., & Jiang, B. (2020). Comorbidities and multi-organ injuries in the treatment of COVID-19. *Lancet, 395*(10228), e52. doi:10.1016/S0140-6736(20)30558-4 PMID:32171074

Zeiler & Fergus. (2014). Visualizing and understanding convolutional networks. In *European Conference on Computer Vision* (vol. 8689). Springer.

Zhang, X. D. (2020). Machine learning. In *A Matrix Algebra Approach to Artificial Intelligence* (pp. 223–440). Springer. doi:10.1007/978-981-15-2770-8_6

ENDNOTE

1 ABC News 2020.

Chapter 12
Privacy Preserving in the Medical Sector:
Techniques and Applications

Nora Alicherif
Sidi Belabess Djilali el Yabess, Algeria

ABSTRACT

A number of fundamental rights are inseparable from human personality. These are in particular the right to the name, the right to missive letters, the right to physical integrity, the moral right of the author and, in the field which interests us, the right to the image, the right to honor, or the "right to secrecy". These rights even have constitutional value. The Constitutional Council considers that they proceed from "respect for individual freedom." In particular, laws that violate the search of vehicles, tax searches, and the publication of certain elements of tax declarations, have been declared unconstitutional. In this chapter the authors will talk about the different techniques that can be used for the preservation of privacy. This chapter will illustrate an overview of the different fundamental techniques for the preservation of privacy in medical sector such as: differential privacy, secure multiparty computation protocol, data anonymization, deidentification, fingerprint privacy, and others.

INTRODUCTION AND PROBLEMATIC

The principle of the protection of private life aims to protect everyone against any form of interference in their private life, it is particularly against revelations made by the press that the legal texts have been put in place. This privacy protection

DOI: 10.4018/978-1-6684-5656-9.ch012

has constantly been threatened and is even more so today, with the increase in the influence of social networks, the decline in moral constraints, and the development of new investigative techniques.

At a time when the world is facing many challenges linked to the development of new technologies, social networks, and the increase in increasingly specialized but also intrusive surveillance measures, it is essential that the balance between the right to security and the right to privacy is maintained and that the protection of privacy remains at the heart of the policies put in place. The right to respect for private life is mentioned very regularly in the opinions of the CNCDH, which recalls its attachment to Article 8 of the European Convention on Human Rights. The right to respect for private life is affirmed by Article 12 of the Universal Declaration of Human Rights of 1948 and codified in French law in article 9 of the Civil Code, which provides that "everyone has the right to respect for his private life."

The state must protect the privacy of its citizens and ensure that invasions of privacy if they do occur, are necessary and proportionate. The CNCDH integrates broad themes regarding private and family life because it is a right that many measures can violate (Yang, 2020).

The next of the chapter will be organised as follows: section two exposes the access and differential privacy techniques, section three details the secure multiparty computation protocols, section four presented different objectives of encrypted privacy preserving, section five we give some algorithms of deidentification, and we finalize with a conclusion and some future works.

ACCESS TO PRIVATE DATA

The increased power and interconnectivity of computer systems available today allow the ability to store and process large amounts of data, resulting in networked information accessible from anywhere. Indeed, thanks to the availability of post-third generation mobile networks, user transactions are no longer bound to the traditional office-centered environment but can be started virtually anywhere and at any hour. Resources may then be accessed in various contexts, and users requesting access may be required to disclose a rich set of distributed information about themselves, including dynamic properties such as their location or communication device and conventional, identity related user attributes. The vast amounts of personal details thus available have led to growing concerns about their users' privacy. Therefore, personal information privacy is an issue that most people are concerned about, mainly because the possibilities of information distribution, combination, and reuse have been increased (Timan, 2021).

In such a scenario, information privacy is about the collection, processing, use, and protection of personal information and should also be addressed by developing privacy aware languages and policies that encompass two notions:

- Guaranteeing the desired level of privacy of information exchanged between different parties and controlling access to services/resources based on this information.
- Managing and storing personal information given to small parties responsibly. A privacy-aware solution should combine these two notions and be expressive and straightforward enough to support the following functionality.

Differential Privacy

Differential privacy is the technology that enables researchers and database analysts to obtain useful information from the databases containing people's personal information without divulging their personal identification. We can describe it as a promise made by a data holder, or curator, to a data subject (owner). The promise is like this: "You will not be affected adversely or otherwise by allowing your data to be used in any study or analysis, no matter what other studies, datasets or information sources are available." This can be achieved by introducing a minimum distraction in the information given by the database. The introduced distraction is immense enough to protect the privacy and, at the same time, limited sufficient so that they provide information to analysts is still valid (Domingo-Ferrer, 2019). Differential privacy can be applied to everything from social networks to location-based services. For example:

- Apple employs differential privacy to accumulate anonymous usage insights from devices like iPhones, iPads and Mac.
- Amazon uses differential privacy to access users' personalized shopping preferences while covering sensitive information regarding their past purchases.
- Facebook uses it to gather behavioral data for target advertising campaigns without defying any nation's privacy policies.

For example, consider an algorithm that analyzes a dataset and computes its statistics such as mean, median, mode, etc. Now, this algorithm can be regarded as differentially private only if via examining the output if a person cannot state whether any individual's data was included in the actual dataset or not.

What Does it Guarantee?

- Differential privacy guarantees mathematically that a person observing the outcome of the private differential analysis will likely produce the same inference about an individual's private information, whether or not that individual's private information is combined in input for the analysis.
- It also specifies verified mathematical assurance of privacy protection counter to a considerable range of privacy attracts such as a differencing attack, linkage attacks, etc.

Secure Multiparty Computation

Secure multi-party computation (also known as secure computation, multi-party computation (MPC) or privacy-preserving computation) provides a cryptographic protocol where no individual can see the other parties data while distributing the data across multi parties. It enables the data scientists and analysts to compute privately on the distributed data without exposing it.

Secure multiparty computation started early in the 1970s. It was known as multiparty computation at that time. It did not gain popularity at that time as it was not implemented practically. In the 1982's, it was introduced as secure two-party multiparty computation. It is used to solve a lot of computation problems without revealing the inputs to other parties. Finally, it came with a name as secure multiparty computation in which the functions of different types are computed; that is why it is sometimes called SFE- Secure Function Evaluation.

- The secure multiparty computation is used to utilize data without compromising privacy.
- It is the cryptographic subfield that helps preserve the privacy of the data.
- Emerging technologies like blockchain, mobile computing, IoT, and cloud computing have resulted in the rebirth of secure multiparty computation.
- Secure multiparty computation has become the hot area of research in the last decade due to the rise of blockchain technology.

Secure Multiparty Computation Protocol

In an MPC, a given number of participants, p1, p2, ..., pN, each have private data, respectively d1, d2, ..., dN. Participants want to compute the value of a public function on that private data: F(d1, d2, ..., dN) while keeping their own inputs secret. For example, suppose we have three parties Alice, Bob and Charlie, with respective inputs x, y and z denoting their salaries. They want to find out the highest of the three salaries, without revealing to each other how much each of them makes. Mathematically, this translates to them computing: F(x, y, z) = max(x, y, z).

If there were some trusted outside party (say, they had a mutual friend Tony who they knew could keep a secret), they could each tell their salary to Tony, and he could compute the maximum and say that number to all of them. The goal of MPC is to design a protocol where, by exchanging messages only with each other, Alice, Bob, and Charlie can still learn F(x, y, z) without revealing who makes what and without having to rely on Tony. They should know no more by engaging in their protocol than they would learn by interacting with an incorruptible, perfectly trustworthy Tony.

In particular, all that the parties can learn is what they can learn from the output and their own input. So, in the above example, if the output is z, then Charlie learns that his z is the maximum value, whereas Alice and Bob learn (if x, y and z are distinct) that their input is not equal to the maximum, and that the maximum held is equal to z. The basic scenario can be easily generalized to where the parties have several inputs and outputs, and the function outputs different values to different parties. Informally speaking, the most fundamental properties that a multi-party computation protocol aims to ensure are:

3.1.1. *Input Privacy*: No information about the private data held by the parties can be inferred from the messages sent during the execution of the protocol. The only information that can be inferred about the private data is what could be inferred from seeing the function's output alone.

3.1.2. *Correctness*: Any proper subset of opposing colluding parties willing to share information or deviate from the instructions during the protocol execution should not be able to force honest parties to output an incorrect result. This correctness goal comes in two flavors: either the honest parties are guaranteed to compute the correct output (a "robust" protocol), or they abort if they find an error (an MPC protocol "with abort") (Monroe, 2021).

Data Anonymization

Data anonymization is the process of protecting private or sensitive information by erasing or encrypting identifiers that connect an individual to stored data. For example, you can run personally identifiable information (PII) such as names, social security numbers, and addresses through a data anonymization process that retains the data but keeps the source anonymous. Data anonymization techniques alter data across systems so it can't be traced back to a specific individual while preserving the data's format and referential integrity. It is one of several approaches organizations can use to comply with stringent data privacy laws that require the protection of personally identifiable information (PII) such as contact information, health records, or financial details (Soni, 2021).

Data Anonymization Techniques

Data masking: hiding data with altered values. You can create a mirror version of a database and apply modification techniques such as character shuffling, encryption, and word or character substitution. For example, you can replace a value character with a symbol like "*" or "x." Data masking makes reverse engineering or detection impossible.

- **Pseudonymization:** a data management and de-identification method that replaces private identifiers with fake identifiers or pseudonyms, for example, replacing the identifier "John Smith" with "Mark Spencer." Pseudonymization preserves statistical accuracy and data integrity, allowing the modified data to be used for training, development, testing, and analytics while protecting data privacy.
- **Generalization:** deliberately removes some of the data to make it less identifiable. Data can be modified into a set of ranges or a broad area with appropriate boundaries. You can remove the house number in an address, but make sure you don't remove the road name. The purpose is to eliminate some of the identifiers while retaining a measure of data accuracy.
- **Data swapping:** also known as shuffling and permutation, is a technique used to rearrange the dataset attribute values, so they don't correspond with the original records. Swapping attributes (columns) that contain identifier values such as date of birth, for example, may have more impact on anonymization than membership type values.
- **Data perturbation:** modifies the original dataset slightly by applying techniques that round numbers and add random noise. The range of values needs to be in proportion to the perturbation. A small base may lead to weak anonymization, while a large base can reduce the dataset's utility. For example, you can use a base of 5 for rounding values like age or house number because it's proportional to the original value. You can multiply a house number by 15, and the value may retain its credence. However, using higher bases like 15 can make the age values seem fake.
- **Synthetic data:** algorithmically manufactured information that has no connection to real events. Synthetic data is used to create artificial datasets instead of altering the original dataset or using it as is and risking privacy and security. The process involves creating statistical models based on patterns

found in the original dataset. You can use standard deviations, medians, linear regression or other statistical techniques to generate the synthetic data.

Example of Data Anonymization in Surveillance Video

Concerning the protection of observed people's privacy, smart video surveillance implies both risk and chance. The systems' increasing capabilities can be misused if we do not deploy suitable counter-measures. On the other hand, when designed with care, such systems can be both privacy-aware and powerful. Smart video surveillance systems aim to assist operators by pointing their attention to certain events that have been detected by intelligent algorithms. From a privacy-protection perspective, video data should only be released in an anonymized fashion, at least until a human operator evaluates whether an event is critical: First-level situation assessment is a matter of what people are doing and not of who people are. The rationale behind privacy-aware smart video surveillance designs is to prevent groundless and unjustified intrusions into people's privacy. In their legal analysis of smart video surveillance according to European data protection law, Bretthauer and Krempel emphasize that people must not be subjected to automated individual decisions made by a system processing personal information. Thus, assessing whether a situation is critical has to be in the hands of a human operator (Troncoso, 2020).

DETECTING AND OBFUSCATING ROIS

Our method for anonymizing videos is as follows. We employ a background estimator for learning the longterm background of the scene as well as for detecting foreground objects, i.e., our regions of interest (RoI) for anonymization. We extract the RoIs from each frame, apply obfuscating image filters or pixel operations, and finally put the anonymized RoIs onto the long-term background model of the scene. With this approach we reduce the risk of privacy breaches due to missing a RoI in some frames. In terms of obfuscating image filters and pixel operations we use pixelization, edge detection, Gaussian blurring, which we apply either on color or on gray-scale RoIs, and reducing the RoI to its silhouette (cf. Fig. 2.2). It also seems natural to combine such operations, e.g., by applying the silhouette technique first and then employ Gaussian blurring afterwards in order to erase smaller structures at the border of the silhouette, which otherwise may reveal visual information about clothes and/or body shapes of persons. However, the focus of our first study is to investigates the effects of each anonymization technique separately (Fan, 2019).

Figure 1. An image Picture taken from a surveillance camera

Figure 2. Anonymization techniques: original pixelation, gray blurring, color gray blurring, edge detection

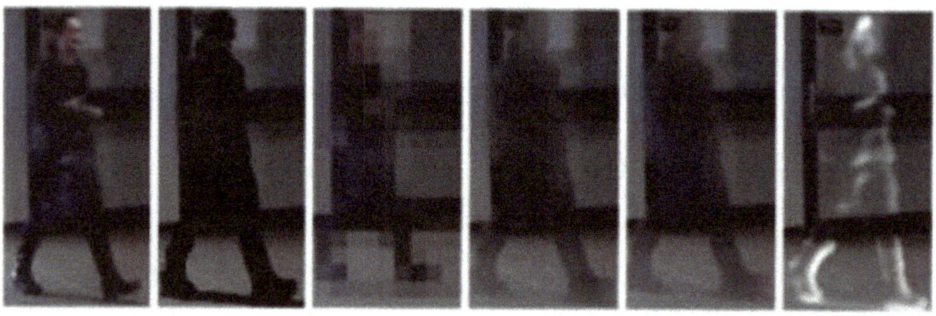

Privacy Preserving Encrypted

Data security is critical not only for businesses but also for home computer users. From client to payment information, personal files, or bank account details, all this information can be hard to replace and potentially dangerous if it falls into the wrong hands. Indeed, data lost due to disasters, such as a flood or fire, is crushing, but losing it because of hackers or malware infection can have much greater consequences. Therefore, data security is the science that studies the methods to protect data in computer and communication systems. It embodies cryptographic controls to face security threats. Cryptography is the science of using mathematics to encrypt and decrypt data to keep messages secured by transforming intelligible data form (plaintext) into an unintelligible form (ciphertext). Every cryptosystem consists of five elements: plaintext, encryption algorithm, decryption algorithm, ciphertext, and Key. The plaintext is messages or data that are in their normal, readable (not encrypted) form. Encryption is the process of converting plaintext to ciphertext by using a key. Ciphertext results from the encryption performed on plaintext using an algorithm called a cipher. Decryption is the process of retrieving the plaintext back from the ciphertext. Finally, the Key uses the information to control the cryptosystem (cipher system), and the sender and the receiver only know it (Lin, 2021).

Cryptography in Data Security

Lawrence Lessig (1999) wrote, "Encryption technologies are the most important technological breakthrough in the last one thousand years." 5 It might seem a slight exaggeration, but it emphasizes the importance of encryption technologies in today's digital world. Indeed, encrypted data play a significant role in protecting data subjects' privacy. Cryptography is the science of using mathematics to encrypt and decrypt data. While cryptography is the science of securing data, cryptanalysis is the science of analyzing and breaking secure communication. Classical cryptography provided secrecy for information sent over channels where eavesdropping and message interception was possible. The sender selected a cipher and an encryption key and either gave it directly to the receiver or else sent it indirectly over a slow but secure channel - typically a trusted courier. Modern cryptography protects data transmitted over high-speed electronic lines or stored in computer systems. Figure 2.3 below shows how a classical information channel works (Leevy, 2020).

ENCRYPTION AND NONLINEAR FUNCTIONS

Nonlinear Functions

In mathematics and science, a nonlinear system is a system in which the output change is not proportional to the change of the input. Nonlinear problems are of interest to cryptographers, engineers, biologists, physicists, mathematicians, and many other scientists because most systems are inherently nonlinear in nature (Ko, 2020).

Figure 3. Classical information channel

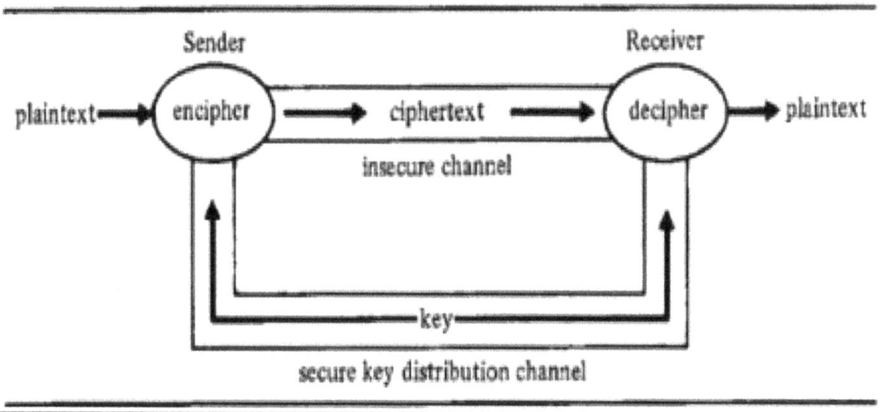

Nonlinear Functions in Cryptography

In the late 1980s, the importance of highly nonlinear functions in cryptography was first discovered by Meier and Staffelbach from the point of view of correlation attacks on stream ciphers and later by Nyberg in the early 1990s after the introduction of the differential cryptanalysis method. Perfect nonlinear (PN) and almost perfect nonlinear (APN) functions, which have the optimal properties for offering resistance against differential cryptanalysis, have since then been an object of intensive study by many mathematicians. In this paper, we survey some of the theoretical results obtained on these functions in the last 25 years. We recall how the links with other mathematical concepts have accelerated the search for PN and APN functions. To illustrate the use of PN and APN functions in practice, we discuss examples of ciphers and their resistance to differential attacks. In particular, we recall that in cryptographic applications, suboptimal functions are often used.

How Linear Function Makes Encryption Strong

Nonlinearity is crucial since most linear systems are easily breakable. As a cipher that explicitly follows the principle of confusion and diffusion, we mention DES. Likewise, this concept applies to other cryptosystems, block ciphers, and stream ciphers.

The theory of perfect nonlinear (or bent) functions has interesting implications for the design of block ciphers as well as stream ciphers. Nonlinearity may not be compatible with other cryptographic design criteria. For example, perfect nonlinearity cannot be achieved in conjunction with balance or the highest nonlinear order. However, a reasonable strategy will be to find nearly perfect nonlinear functions which satisfy additional design criteria. This is illustrated by the following example of finding almost perfect nonlinear functions which are balanced.

If you combine functions linear of the same field you get again a function linear over that field, and linear equations are easy to solve. So, you do need something non-linear to make solving difficult.

Example of Nonlinear Function in AES

The non-linear operation is AES's S-box, which is a finite-field inverse S(x) = x-1. You can see that the S-box is non-linear because it is not necessarily true that $(x+y)^{-1} = x^{-1}+y^{-1}$

DATA AUGMENTATION AND PRIVACY

Data augmentation is a strategy that enables practitioners to significantly increase the diversity of data available for training models, without actually collecting new data. Data augmentation techniques such as cropping, padding, and horizontal flipping are commonly used to train large neural networks (Kumar, 2019) . Data augmentation is often necessary when confronted with a massive domain of use like an open world as it improves the robustness of a neural network. Data augmentation is useful but can be time-consuming thus slowing down your development. In order to scale down the cost of training, you need adequate data augmentation (Kumar, 2019).

Data Augmentation in Privacy

Deep learning models often raise privacy concerns as they leak information about their training data. This leakage enables membership inference attacks (MIA) that can identify whether a data point was in a model's training set. Research shows that some 'data augmentation' mechanisms may reduce the risk by combatting a key factor increasing the leakage, overfitting. While many mechanisms exist, their effectiveness against MIAs and privacy properties have not been studied systematically. Employing two recent MIAs, we explore the lower bound on the risk in the absence of formal upper bounds. First, we evaluate 7 mechanisms and differential privacy, on three image classification tasks. We find that applying augmentation to increase the model's utility does not mitigate the risk and protection comes with a utility penalty. Further, we also investigate why popular label smoothing mechanism consistently amplifies the risk. Finally, we propose 'loss-rank-correlation' (LRC) metric to assess how similar the effects of different mechanisms are. This, for example, reveals the similarity of applying high-intensity augmentation against MIAs to simply reducing the training time. Our findings emphasize the utility-privacy trade-off and provide practical guidelines on using augmentation to manage the trade-off (Azbeg, 2022).

The Way it Works

Computer vision applications use common data augmentation methods for training data. There are classic and advanced techniques in data augmentation for image recognition and natural language processing (NLP).

Figure 4. Data augmentation methods

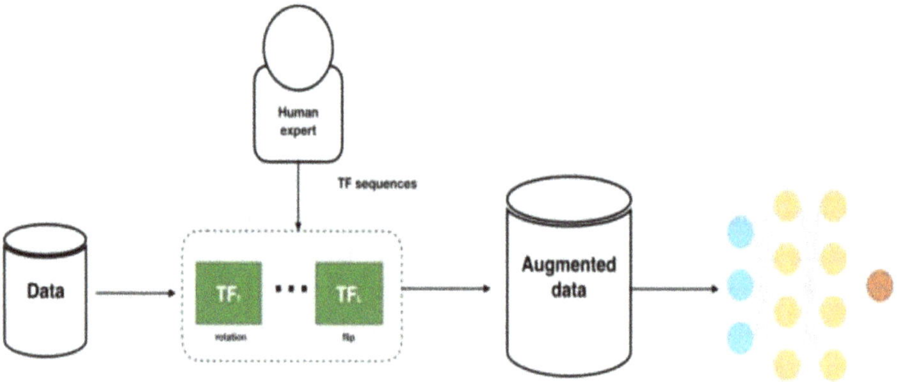

DEIDENTIFICATION

Data de-identification refers to breaking the link between data and the individual with whom the data is initially associated. Essentially, this requires removing or transforming personal identifiers. Once personal identifiers are removed or transformed using the data de-identification process, it is much easier to reuse and share the data with third parties. Data de-identification is expressly governed under HIPAA, which is why most people associate the data de-identification process with medical data. However, data deidentification is also important for businesses or agencies that want or need to mask identities under other frameworks, such as CCPA and CPRA, or even GDPR (Cohen, 2020).

When applied to metadata or general data about identification, the process is also known as data anonymization. Common strategies include deleting or masking personal identifiers, such as personal name, and suppressing or generalizing quasi-identifiers, such as date of birth. The reverse process of using de-identified data to identify individuals is known as data re-identification. Successful re-identifications cast doubt on deidentification's effectiveness. A systematic review of fourteen distinct re-identification attacks found "a high re-identification rate dominated by small-scale studies on data that was not de-identified according to existing standards".

7.1. **HIPAA:** names two different methods of de-identifying data: Safe Harbor and Expert Determination.

- Safe Harbor: The Safe Harbor method of de-identification requires removing 18 types of identifiers, like those listed below, so that residual information cannot be used for identification:
- Names
- Dates, except the year
- Telephone numbers
- Geographic data
- Fax numbers
- Social Security numbers
- Email addresses
- Medical record numbers
- Account numbers
- Health plan beneficiary numbers
- Certificate/license numbers
- Vehicle identifiers and serial numbers, including license plates
- Web URLs
- Device identifiers and serial numbers
- Internet protocol addresses
- Full face photos and comparable images
- Biometric identifiers
- Any unique identifying number, characteristic, or code

Any of these identifiers can classify health information as protected health information (PHI), which limits use and disclosure. Data de-identification tools with sensitive data discovery can detect and mask such information. The Safe Harbor method, which is usually praised for its simplicity and low cost, is not well adapted for all use cases: It can either be overly restrictive, leaving too little utility within the data, or overly permissive, leaving too many indirect identifiers in the clear.

- **Expert Determination:**Expert determination involves applying statistical and scientific principles to data to achieve a very small risk of re-identification.

Figure 5. Two methods to achieve de-identification in accordance with the HIPAA Privacy Rule (Zhang, 2018)

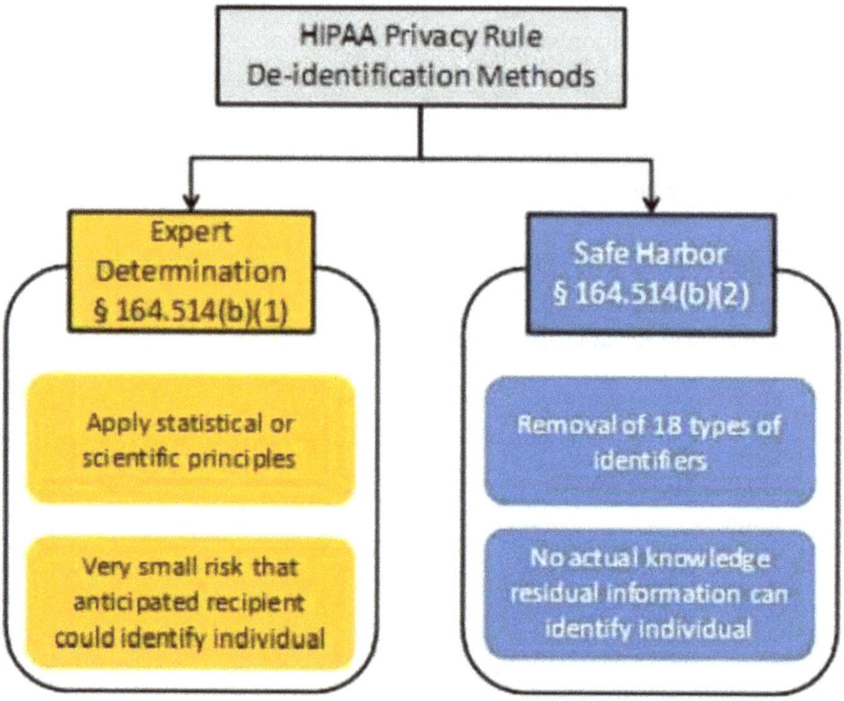

> This method makes it possible to tailor the de-identification process to the use case at hand while also maximizing utility; it is therefore praised for its flexibility.

Expert determination is sometimes considered too costly to use because it requires the involvement of an expert in statistics, who can be expensive to source. However, the expert determination method enables the use of quantitative methods to lower the re-identification risk, which opens the door for leveraging generalization and automation.

TECHNIQUES OF DE-IDENTIFICATION

Common strategies of de-identification are masking personal identifiers and generalizing quasi-identifiers. Pseudonymization is the main technique used to mask personal identifiers from data records, and k-anonymization is usually adopted for generalizing quasi-identifiers:

Pseudonymization

Pseudonymization is performed by replacing real names with a temporary ID. It deletes or masks personal identifiers to make individuals unidentified. This method makes it possible to track the individual's record over time even though the record will be updated. However, it cannot prevent the individual from being identified if some specific combinations of attributes in the data record indirectly identify them (Shuaib, 2021).

8.2. **Generalizing (k-anonymization)** K-anonymization is a data generalization technique that is implemented once direct identifiers have been masked. The k-anonymization process reduces re-identification risks by hiding individuals in groups and suppressing indirect identifiers for groups smaller than a predetermined number, k. This aims to mitigate identity and relational inference attacks. This de-identification technique can help reduce the need for data redaction in data sets, which helps increase its utility without compromising data privacy.

8.3. **Randomizing** (differential privacy and randomized response) Differential privacy is a randomization technique that is implemented once direct identifiers have been masked. There are two approaches to differential privacy: local and global (Hassan, 2019).

Local differential privacy is a data randomization method that usually applies to sensitive attributes and offers a mathematical guarantee against attribute-based inference attacks. This is accomplished by randomizing attribute values in a way that limits the amount of personal information inferable by an attacker while still preserving some analytic utility, since gathering too much information on a specific record can undermine privacy. Individuals whose data is included in the queried data set are therefore able to deny the specific attributes attached to their records. Technology companies like Google and Apple, which collect a wide range and huge amount of personal data, have adopted local differential privacy.

Global differential privacy is a method that randomizes aggregate data. This approach constrains data users to only formulate aggregate queries (e.g., count, average, max, min, etc.), and offers a mathematical guarantee against identity-, attribute-, participation-, and relational-based inference attacks. Individuals whose data is included in the queried data set are therefore able to deny their participation in the data set as a whole. The US Census Bureau, for example, employs global differential privacy because aggregation on its own is insufficient to preserve privacy (Niu, 2021).

FINGERPRINT PRIVACY

Prints are the impression made by the papillary ridges on the ends of the fingers and thumbs. Fingerprints afford an infallible means of personal identification because the ridge arrangement on every finger of every human being is unique and does not alter with growth or age. Fingerprints serve to reveal an individual's true identity despite personal denial, assumed names, or changes in personal appearance resulting from age, disease, plastic surgery, or accident. Utilizing fingerprints as a means of identification, referred to as dactyloscopy, is an indispensable aid to modern law enforcement (Azbeg,2022).

Each ridge of the epidermis (outer skin) is dotted with sweat pores for its entire length and is anchored to the dermis (inner skin) by a double row of peglike protuberances, or papillae.

Any ridged area of the hand or foot may be used as identification. However, finger impressions are preferred to those from other parts of the body because they can be taken with a minimum of time and effort. The ridges in such impressions form patterns (distinctive outlines or shapes) that can be readily sorted into groups for ease in the filing.

Fingerprints are classified in a three-way process: by the shapes and contours of individual patterns, by noting the finger positions of the pattern types, and by relative size, determined by counting the ridges in loops and by tracing the ridges in whorls. The information obtained in this way is incorporated into a concise formula, which is known as the individual's fingerprint classification (Sowmiya, 2021).

Fingerprint Identification

A person's fingerprint is compared to stored fingerprint data in fingerprint identification. A fingerprint can, for example, be stored in an identification system database, a passport chip, or an access card's memory. Identification locations include a door side fingerprint reader, a reader connected to a computer, or a fingerprint reader integrated into a smartphone. Fingerprint identification is nearly always combined as a part of another system, such as a locking system. Fingerprint identification is used to verify a person's identity, after which the system can perform the required actions, such as opening a lock, allowing a person to use software, or enabling a machine to start (Valdes-Ramirez, 2019).

Fingerprint Identification Implementation

Fingerprint identification is based on pattern recognition, where the arches, loops and whorls of the fingerprint ridges are compared with stored data. Identification is performed in three parts:

- A picture is taken of the fingerprint. The image can be taken optically with a camera in the reader or electronically or combined with these two methods. The end result is a digital black and white photograph of the ridges in the fingerprint.
- The fingerprint is then transformed into a numerical model which stores the fingerprint's unique characteristics, such as the arches and loops and their distance from each other, as a series of numbers.
- A recognized numerical model is compared with a stored numerical model (or models) to find similarities.

CONCLUSION

In this chapter, we have tried to highlight the most important means and techniques of privacy preservation, encrypted data anonymization, and fingerprint used by researchers to preserve information of all kinds and prevent its spread illegally and how to preserve an individual's private life.

REFERENCES

Azbeg, K., Ouchetto, O., & Andaloussi, S. J. (2022). *Access Control and Privacy-Preserving Blockchain-Based System for Diseases Management*. IEEE Transactions on Computational Social Systems.

Cohen, D. (2020). HIPAA Reform or a Patchwork Scheme: A Look at Preemption, Scope, and the Inclusion of a Private Right of Action in a New Federal Data Privacy Law. *American University Washington College of Law*.

Domingo-Ferrer, J., Farras, O., Ribes-González, J., & Sánchez, D. (2019). Privacy-preserving cloud computing on sensitive data: A survey of methods, products and challenges. *Computer Communications*, *140*, 38–60. doi:10.1016/j.comcom.2019.04.011

Fan, L. (2019, July). Practical image obfuscation with provable privacy. In IEEE international conference on multimedia and expo (ICME) (pp. 784-789). IEEE. doi:10.1109/ICME.2019.00140

Hassan, M. U., Rehmani, M. H., & Chen, J. (2019). Differential privacy techniques for cyber physical systems: A survey. *IEEE Communications Surveys and Tutorials*, *22*(1), 746–789. doi:10.1109/COMST.2019.2944748

Karunarathne, S. M., Saxena, N., & Khan, M. K. (2021). Security and privacy in IoT smart healthcare. *IEEE Internet Computing*, *25*(4), 37–48. doi:10.1109/MIC.2021.3051675

Ko, D. H., Choi, S. H., Shin, J. M., Liu, P., & Choi, Y. H. (2020). Structural image de-identification for privacy-preserving deep learning. *IEEE Access: Practical Innovations, Open Solutions*, *8*, 119848–119862. doi:10.1109/ACCESS.2020.3005911

Kumar, A., & Kumar, R. (2019, March). Privacy-Preservation of Vertically Partitioned Electronic Health Record using Perturbation Methods. In *6th International Conference on Computing for Sustainable Global Development (INDIACom)* (pp. 161-166). IEEE.

Leevy, J. L., Khoshgoftaar, T. M., & Villanustre, F. (2020). Survey on RNN and CRF models for de-identification of medical free text. *Journal of Big Data*, *7*(1), 1–22. doi:10.118640537-020-00351-4

Lin, J., Li, Y., & Yang, G. (2021). FPGAN: Face de-identification method with generative adversarial networks for social robots. *Neural Networks*, *133*, 132–147. doi:10.1016/j.neunet.2020.09.001 PMID:33217682

Monroe, C., Tazi, F., & Das, S. (2021). Location data and COVID-19 contact tracing: how data privacy regulations and cell service providers work in tandem. *Researchgate*.

Niu, X., Song, T., & Xiong, H. (2021). Large scale synthesis of red emissive carbon dots powder by solid state reaction for fingerprint identification. *Chinese Chemical Letters*, *32*(6), 1953–1956. doi:10.1016/j.cclet.2021.01.006

Shuaib, M., Alam, S., Alam, M. S., & Nasir, M. S. (2021). Compliance with HIPAA and GDPR in blockchain-based electronic health record. *Materials Today: Proceedings*. doi:10.1016/j.matpr.2021.03.059

Soni, M., & Singh, D. K. (2021). Privacy-preserving authentication and Key-Management protocol for health information systems. In *Data Protection and Privacy in Healthcare* (pp. 37–50). CRC Press. doi:10.1201/9781003048848-3

Sowmiya, B., & Poovammal, E. (2021). A heuristic K-anonymity based privacy preserving for student management hyperledger fabric blockchain. *Wireless Personal Communications*, 1–18.

Timan, T., & Mann, Z. (2021). Data Protection in the Era of Artificial Intelligence: Trends, Existing Solutions and Recommendations for Privacy-Preserving Technologies. In *The Elements of Big Data Value* (pp. 153–175). Springer. doi:10.1007/978-3-030-68176-0_7

Troncoso, C., Payer, M., Hubaux, J. P., Salathé, M., Larus, J., Bugnion, E., & Pereira, J. (2020). Decentralized privacy-preserving proximity tracing.

Valdes-Ramirez, D., Medina-Pérez, M. A., Monroy, R., Loyola-González, O., Rodríguez-Ruiz, J., Morales, A., & Herrera, F. (2019). A Review of Fingerprint Feature Representations and Their Applications for Latent Fingerprint Identification: Trends and Evaluation. *IEEE Access: Practical Innovations, Open Solutions*, *7*(1), 48484–48499. doi:10.1109/ACCESS.2019.2909497

Yang, M., Lyu, L., Zhao, J., Zhu, T., & Lam, K. Y. (2020). Local differential privacy and its applications: A comprehensive survey. *arXiv preprint arXiv:2008.03686*.

Zhang, Z., Guo, X., & Lin, Y. (2018). Trust management method of D2D communication based on RF fingerprint identification. *IEEE Access: Practical Innovations, Open Solutions*, *6*, 66082–66087. doi:10.1109/ACCESS.2018.2878595

Chapter 13
Technological Innovation Driven by Big Data

G. Taviti Naidu
Dr. L. Bullayya College, India

K. V. B. Ganesh
KLEF Vaddeswaram, India

V. Vidya Chellam
Madurai Kamaraj University, India

S. Praveenkumar
https://orcid.org/0000-0001-9639-5817
Madurai Kamaraj University, India

Dharmesh Dhabliya
Vishwakarma Institute of Information Technology, India

Sabyasachi Pramanik
https://orcid.org/0000-0002-9431-8751
Haldia Institute of Technology, India

Ankur Gupta
https://orcid.org/0000-0002-4651-5830
Vaish College of Engineering, India

ABSTRACT

Large-scale research evaluations, as well as the acquisition and analysis of massive amounts of data, in order to follow trends and models form the organization. As the world around us continues to change, many entrepreneurs are interested in learning how big data might help them enhance their businesses. While businesses have traditionally analyzed data, recent technical advancements have spawned new entrants and unlocked the potential of big data. The extent of the consequences of big data, according to an SNS research, is presently $57 billion and expected to rise further.

DOI: 10.4018/978-1-6684-5656-9.ch013

INTRODUCTION

What role does big data play in driving business trends, and what does the future hold for the industry? Big data need a highly trained staff. Smart firms utilize big data to increase consumer engagement, develop new markets for their marketing, and improve operational performance. However, if a company has the human resources required to take benefit of this cutting-edge contemporary technology, it will be unable to do so. Companies are continuously pursuing a thorough grasp of data collection and assessment companies. These expert data masters gather up the companies that hope to plant the Fortune 500 one day, with some companies also planning to guarantee that their personnel are adequate. While employees with a university education are presently highly valued, the workplace of the future will need people with more sophisticated abilities and technology.

Consider Teradata's 2017 Development Survey, which found that over half of global companies are experiencing a shortage of technology-skilled people. Businesses will want more imaginative employees who are not intimidated by developing technologies as the big data sector grows. As big data becomes more spectacular, companies need ensure that their employees are eager to analyze it and make better judgments. Due to the absence of productivity of workers, every creative company's leftovers are the polar opposite.

Because of big data's revolutionary power, a range of major markets have been transformed. Algorithms are expected to heavily govern the finance sector in the next years, smashing unusually large-scale data endeavors to anticipate industry changes and discover flaws in financial operations (Welc, 2022). Big data (Schloemer, 2022) is also propelling emerging industries like financial processing, driverless cars, and smart homes (Nyangaresi, 2022) forward. To stay a keep away from collisions and modify their routes, tomorrow's incredible smart automobiles will rely on gathering and analyzing locally specified data. Even digital data is responsible for the market position of many established firms. Big data is a term that refers to a large Netflix, a well-known website, has now become self-evident, and basically re-forming the home video industry by capturing and analyzing user data. By continually upgrading data stages, the organization will determine which of its shows is better in particular random markets, predict which pilots it can finance, and even predict the amount of leisure grants it will get.

For large organizations, utilizing and understanding big data is a crucial competing edge. Organizations can have a possibility to know about initial vision that their opponents don't know whether they can acquire extra data from their present clients and technologies.

Various opportunities for latest advancements may be provided by big data, from internal perceptions to direct client commitments. The utilization of technologies, extensive investigation, and data-driven analysis are 3 important commercial potential.

AUTOMATION

By robotic process automation, big data has the capability to intensify competence and functioning in an organization. For automated decision-taking, large volumes of real-time data can be processed fast and consolidated into corporate functions. Computerizing data collection and storage is achievable due to scalable IT framework and reduction of cloud computing costs.

Detailed Insights

Big data can be utilized to expose chances that organizations were ignorant of earlier possessing the ability to evaluate large volumes of data. Also, the forming of newer or enhanced elements can be done utilizing composite data sets. Market-dependent proprietary data is useful in a cutthroat atmosphere.

Superior and Quicker Decision-Making

Organizations may now rapidly evaluate knowledge and take educated, informed choices and courtesy to the speed of data analysis frameworks and the enhancements to survey newer origin of data.

How to Use Big Data to Your Advantage

The necessity for big data analysis is extensive; since 2015, more than 55% of major firms made investments in big data technology. To handle the many potential data points, it might be difficult to know where to start.

The following factors should be taken into account by enterprises before choosing and establishing a big data solution.

Expertise in a Big Data Team

Establish a group of experts in data gathering, analysis, and technique to aid in developing the perfect big data plan that generates profits for the business. People who are knowledgeable about current analytical techniques, adept at working with huge volume of data sets, and specialists who understand broad business objectives should be on this team.

Select Final Objectives

For a big data solution to be implemented effectively, having the appropriate goals is essential. The organization's ultimate objectives must be correctly matched with the data and analytics (i.e., huge gain, brand acknowledgement, sales figures).

Obtain Accurate Data

The next stage after exhibiting the major commercial goals is in thoroughly comprehending the data prior to implementing it. The foundation of the whole big data methodology is recognizing, obtaining, and tracking the appropriate data. Utilizing the erroneous data sets may have adverse effects and steer the whole business in an incorrect way.

Use the Right Analytical Techniques

Professional data analysts are adept at precisely and quickly turning massive data into insightful information. Aiding squads process data analytics and build swift marketing choices more quickly when the data is presented in digestible, visual summaries.

TOOLS FOR BIG DATA ANALYTICS

Big data's emergence has given birth to a sizable market for data analytics tools that assist businesses in smoothly implementing big data solutions. Startups with expertise working fast with major organizations to prepare data for action include Ople and Cruz Informatics.

Ople was established with the clear goal of making artificial intelligence simple, affordable, and commonplace. Ople quickens the data science methodology, allowing businesses to include more problems and find solutions more quickly. Data scientists can create ten times more manufacturing-dependent Artificial Intelligence techniques with Ople and cut the span of stationing from years to a lesser number of days by concentrating on the business intelligence.

Figure 1. Big data functionalities in an enterprise

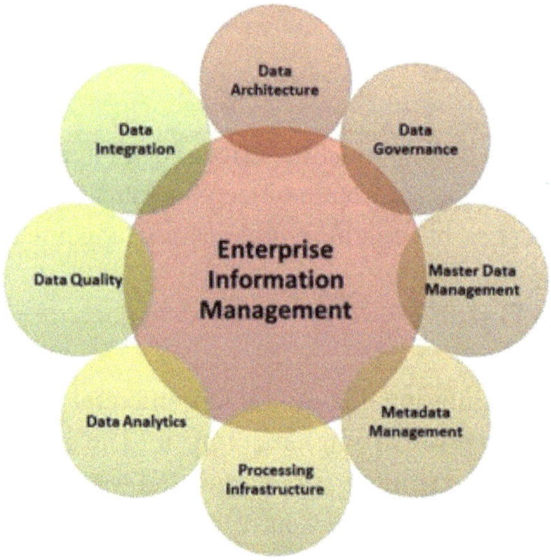

Informatics Crux

An industry's lifeblood is the use of data to unearth useful insights. Crux helps businesses easily obtain, examine, and alter data so they can concentrate on what truly matters by taking on the cumbersome portions of their information supply chains. This makes data actionable and ready for action. Crux provides a safe, scalable environment to store, analyze, and transform data via its built-in cloud computing approaches with its Informatics Platform. Data is made enjoyable with Crux.

Big Data as A Tool for Business Success

If more firms and government agencies see the benefits of big data, there's little question that more money will be invested in it to put it to good use. If insurance companies want to determine who among their clients is responsible for injuries, they will utilize increasingly sophisticated technologies to detect hazard. Apple and Google, for example, will research how their new devices will be offered to their current customers. Big data has nearly limitless ramifications.

The success report from IBM focuses on exploiting large-scale data to identify process flaws and draw consumer expectations about the things they purchase, such as 3D (Goh et. al., 2022) and wearable technology (Madanian et. al., 2021). IBM's innovations are built on one of the most widely used technologies.

How CEOs and CIOs early participate in inquiry to assist manages their firms and forecast the future is one of the most important factors for a company's success.

Businesses may need more usable data sources for interpretation and analysis as the internet becomes more sophisticated. There is an entire sector dedicated to companies that are sufficiently cautious to participate in scientific forecasting.

With computerization and the ability to challenge present economies, views for the destiny of the digital world are constantly concerning, but they seldom attain their full potential for significant effects. In the near future, this extraordinary method will be utilized to make, market, and deliver almost any item and service.

Individuals and enterprises that do not want to be left alone should learn about the destiny of the knowledge sector and make plans for it. There's a whole new future, and it's one that revolves on big data.

Without a question, there have been significant advancements in big data and business intelligence (Kanan et. al., 2022). They have a tremendous power to accelerate social and economic benefits, ranging from human well-being to sustainable urban life. Using such gadgets to incapacitate a person, on the other hand, is absolutely improper. Personal data collected centrally for regulated monitoring was manipulated in an authoritarian manner by heavy pushing and resident evaluations. It is not only contrary to human rights and democratic norms, but it is also unsuitable for leading new, innovative societies. Enhanced intelligence and risk mitigation strategies are plainly required to solve the world's genuine concerns. The Data for Humanity program (see "Digital Data for Society and Civilization") and responsible software studies provide guidance on how Digital Data and Business Analytics (Zamani et. al., 2022) might be utilized to benefit society.

Big Data's Impact on Disruptive Innovation

A large majority of individuals approve, despite the different implementation variants. The term "disruptive development" refers to the following:

- (From a client's perspective) low-cost
- Technology that is easier to use or that can be distributed
- There are more options available.
- Use an auxiliary factor model as well (relative to current arrangements)

The reality that each of the three companies has a hard time adjusting to competition is one of the main reasons why these disruptive traits are important. When one or two pieces of an infrastructure, highly qualified individuals, or an out-of-date logistics system become obsolete, it is difficult to adjust quickly to current circumstances. Hundreds of people being laid off, consumers being inconvenienced, and billions of dollars in investment – executives find it difficult to examine these factors and take real-world realities into consideration.

We sought for signals of impending disruptions in the low end of the scale, and we found them. Innovative products looked to be more cramming than existing high-end structures due to their smaller, more open, and centered on new technological designs. Because of their lower prices, they were able to attract purchasers who had earlier been turned down. Apple initially built a gadget that was affordable enough for adolescents to be interested in, customers who would not have considered purchasing a DEC device. Their cost advantage is advantageous to them. Sony is well-known for its "compactness" in television. Before the transistor, no one realized you could do it fairly. Emerging breakthroughs, in connection with the evolution of a business strategy, have the basic economic benefit necessary to capture large sectors of the industry over time.

When you return to the previous description, however, the concept that reduced entry was accomplished from a troublesome method mill isn't at the forefront. Instead, it was a byproduct. Rather. What is the explanation behind this? Because a newcomer cannot provide better value in a competitive market that has been polished over decades, goods however, although the low-end approach was popular, it was not what kept competitors in business. They were driving marginal profit increases via their own pricing policies and focus, effectively keeping these companies out of the marketplace. CEOs are condemned to battle as long as the current choice (trying to make obsolete facilities more useful) maintains the off-base option on a long-term posture (don't take up new technology).

Unfortunately, the focus on low-end disruption just obscures our capacity to see the things: lower, more open, and costly. In particular, data-driven (Dushyant et. al., 2022) industry scientists are likewise perplexed about enabled interrupters. To get a feel of this, look at a few hotly contested instances.

Is super upsetting to you? 'No, since their initial offering was at the upper end of the market,' says the outsider. The correct approach is to acknowledge that the stage they provided ultimately enabled them to have lower-cost drivers (like UberX) and give more affordable, open-air transportation options, which provide a fundamental economic advantage for both taxi companies and, most likely, automobile ownership. The software's simplicity is by far the easiest and most obvious feature to duplicate.

Was the transition from Google's to Nokia's cell phones disruptive? "No," they say, "since their core smart phones were more useful than Nokia's own phones, which dominated the worldwide scene." The right response is that the strategy of creating an atmosphere of utilization development on stage enabled them to operate in a more comprehensive way, which is (all) cheaper.

Is 23andMe posing a danger to pharmaceutical companies? "No, since it is in two independent verticals," the foundation responds. One is found in our forefathers, while the other is found in pharmaceutical manufacturing. The argument is that 23 of us have vast amounts of data that will enable them to begin developing treatments that are less expensive, more accessible, and intrinsically useful. That is the correct response.

In each of these circumstances, the end outcome is disruption. Both of these circumstances encourage current managers to avoid the stumbling block and stick with their present solutions. Taxi companies attempted to take advantage of laws in order to maintain the reputation of their medallion and drivers. Nokia's attempt to safeguard its closed environment and keep thousands of Symbian-oriented employees employed was foolish. And you can be guaranteed that Merck, Pfizer, and Roche have better advantages than 23andMe, while choosing a totally different strategy to making the most of their high-end R&D businesses.

In the subject of development, there is something fresh that is missing from the conversation. New means of addressing markets are accessible in different ways at the centre of most recent revolution. Uber built a stage in the competitive limo industry that allows for more reliable transportation and organization. Netflix stole your eyes and used the data required to blast the content development process in streaming videos. Google mapped the whole planet before learning how to build self-sufficient traffic and street forms.

There is no doubt that this is a breach of privacy (Bandyopadhyay et. al., 2021). These companies create more affordable and useful items than their competitors. It does not, however, begin at the bottom end of the market; rather, it emerges from symmetrical sectors with significant knowledge synergy. It all starts with the data base, and then the information infrastructure may be organized to target an existing market.

Administrators, engineers, and entrepreneurs must stop disputing whether classical disruption satisfies the requirements. The disruption that data may cause is unique to the paradigm shift, but it's already here—and it's here to stay. Uber's devastation of the taxi business is an illustration of the present system's extraordinary cost benefit and lack of response. The following should be the new issues:

- "How do you think you'd be able to adapt to this new form of competition?"
- "How do you assess new threats?" says the narrator.
- "When data is a fundamental element of every new disruption, what skills do you need and where do you acquire them?"

Challenged firms have an excellent method of seeing future risks in conjunction with a desire to maintain long-term commitments protected – via transformational revenue – in order to survive in this shifting atmosphere. So, how are we going to proceed now? First and first, people's basic rights should be protected, especially in this era of digital transformation, since they are the foundations of a modern, functioning, democratic society. This necessitates the establishment of a new mutual trust and cooperation arrangement in which individuals and customers regard themselves as allies rather than obstacles or instruments to be used. For this to occur, the state must have an efficient legislative apparatus in place to ensure that technology is created and applied in democratically acceptable ways. This should provide substantial material self-determination, as well as other aspects of self-determination and accountability in our lives. This is a prerequisite.

Even a replica of personally identifiable information obtained about us should be safeguarded. It should be controlled by law in order to address their data consumption (perhaps with the help of unique AI-based digital partners), and the like information would have to be sent to a centralized personal data storage place. To guarantee better protection and discourage separation, illicit data usage should be forbidden by law. People will then decide who should utilize their knowledge for what cause and to what extent. Furthermore, necessary actions should be made to guarantee that data is stored and shared in a secure manner.

The accuracy of the information on which we make our judgments will be substantially improved by advanced credibility systems with multiple characteristics. We may perceive issues from numerous distinct perspectives and are less likely to control them by skewed information if the user selects and customizes data filters (Koren et. al. 2022), search recommendations (Argent et. al., 2022), and estimations.

We also demand an effective public grumbling procedure, as well as successful acquiescence for law violations. To achieve enough accountability and trust, premier research organizations should act as stewards of data and computations that now elude democratic supervision. This also necessitates a reasonable, exact code that everybody with access to sensitive data and statistics ought to be able to follow, as a form of Hippocratic pledge for IT professionals.

Furthermore, we will need a worldwide strategy that includes the establishment of a foundation for rising businesses as well as the development of civic culture. Every year, we spend billions of dollars on agriculture (Pramanik et. al., 2022), public infrastructure, schools, and universities.

So, what public structures are required for contemporary civilization to succeed? To begin with, completely new educational concepts are required. Rather of developing automated labor, the focus should be on critical thinking, creativity, innovation, and entrepreneurship (whose duties are to be carried out by machines and computer calculations in the future). People must be aware of the connectivity between the developing world and the physical world, therefore education should include knowledge of how to utilize new technology effectively and practically. People must understand these approaches, as well as which uses are unlawful, in order to defend their rights in a safe and reasonable manner. It is all the more critical, then, to make information available to research, industry, government, and educational organizations.

Self-employment, the establishment of their own businesses, the investigation of coordinate partners, the manufacture of global market products and services, capital management, and the supply of a structured savings pledge at government cost all need a participatory stage (a form of shared economy for all). Furthermore, municipalities and villages may establish new digitals centers – for example, amazing labs – where ideas may be generated and tested for free. The openness and creativity with which these centers operate might help to bring about tremendous social transformation.

Specialized competitions may give additional impetus for change, increase civic exposure, and act as a catalyst for a more active digital community. It may be particularly beneficial in bringing together common groups in order to guarantee that their responses to global concerns are as near as possible (e.g. "Climate Olympics"). Etapes aimed at combining limited capital, for example, might aid in unlocking the enormous potential of the mostly untapped circular and social economy.

With a commitment to an open data strategy, governments and business will increasingly disclose data to scientists and the general public, creating the necessary circumstances for effective and creative knowledge and an environment that fulfils our needs. Tax breaks, such as those offered in certain nations to encourage the adoption of renewable technologies, may also help.

Third, building a "human-controlled digital nervous system" might open up new Internet of Things (Pramanik, 2022) opportunities for everyone and give real-time data measurements. The positive and negative adverse effects of our connections with people and our ecology must be examined in order to enable effective energy usage and minimize climate change. Processes may be steered by self-association to create the desired outcomes when using effective feedback rings.

For successful adoption, we'll need different incentives and sharing systems that are available to all technical, political, and societal trend-setters. This will provide fresh new possibilities and, as a result, greater growth. Modern pay laws and a pluralistic financial system (e.g., fully different foreign currencies) will greatly support the digital economy's boundless potential for all purposes.

To deal with the volatility and distinctiveness of our future and convert it into an advantage, we'd need personal digital employees. Computerized logic advancements might likewise assist these automated coworkers. Numerous networks aggregating individual and citizen awareness are expected to be freely established and altered as needed in the future. These networks, on the other hand, should be spread with the ultimate goal of maintaining control over our lives. In fact, you should be able to log in and leave whenever you want.

THROUGH BIG DATA, REIMAGINING BUSINESS INNOVATION, COMPETITIVENESS, AND PRODUCTIVITY

Big data has penetrated a wide range of industries and is now seen as a genuine success driver, a cost advantage, and an increase in efficiency.

Big Data is at the centre of a technological revolution that has reached the world's most vulnerable reaches. The concept is defined by the 5 Vs: number, variation, distance, actual value, and, most importantly, meaning. This final V is why Big Data has become a partner in industries like manufacturing, retail, and healthcare, as well as more conventional domains like government and even extortion and crime detection. In a digitally oriented market, Big Data gradually becomes a cause of innovation and money, as well as a tactic for gaining a strategic edge.

- **Catalyst for Innovation**

Advancement may be able to address several crucial issues, such as R&D costs, customer acceptance, and the transition from early investors to the mainstream market.

- **Create New Products**

The foundation for studying and nurturing emergent patterns and ideas will be big data. A corporation may identify unmet demands, niches that aren't well-established, and categorize marketing operations for the correct audience using basic research equations.

The utilization of big data might assist the product in any stage of development, from conception to delivery. The novel ideas are centered on online networking, corporate news may recognize the necessity for the establishment of an open door for a specific project, and existing databases can validate the concept and assist in the definition of the template.

- **Developing and Changing Model Businesses**

When data becomes a commodity in and of itself, organizations may leverage the information gathered for their existing operations to create or enhance their value propositions. The very same sets may be valuable for competitors, comparable businesses, and other organizations with similar consumer bases. Could a corporation with a large data base set up a subsidiary and generate many revenue streams? For example, a pharmacy clerk who keeps track of prescriptions dispensed may sell his information to a health-research organization or even an insurance company.

- **Increasing your Competitiveness**

If a company's golden customer group can be identified, its interests better served, and its expenses are perfectly matched to its revenues, it can function to its maximum potential. To improve speed, Big Data may change any of these variables.

- **Creating Enough Segmentation**

For patterns and cluster construction, big data computational knowledge streams are used. Clients may be grouped into related sessions using this way. According to an expert from Indata Labs Big Data Counsels, "by breaking down huge measurements of seemingly unrelated information, calculations that detect trends and affiliations and build customer segments separate yet pertinent for their behavior on the brand or product from demographics or social data only". Facebook used the phrase "lookalike crowds" to describe a tool for finding similar consumer meets of companies that are currently succeeding.

- **Reputation on the Internet**

Against competitors, a sentiment analysis of information eagerly transmitted by consumers or admin in web-based networking sites should be available. Checking your company's comprehension, its capabilities, and their wants against the products of competitors will add to new business. Scratching your brand's online environment may also reveal any negative comments or even unscrupulous practices that might affect your business.

- **Enhancement of the User Experience**

I can only get a glimpse of who the clientele are and what they think of you by looking at a larger picture. The genuine difficulty comes when firms seek to make their customers happy than their competitors. You'll study about the game both online and offline, then adjust your bargain to their preferences based on what you've learned. Cookies, external data, and census data are all permitted in this struggle. Customers are aware of how firms collect information, and they want high quality and low costs in return, so they are hesitant to give up any of their autonomy.

- **Increasing Productivity Saving Time**

Employees who keep track of their everyday routines are inefficient, but the selecting and decision-making procedures that are based on their analysis will certainly result in double-digit growth. During work sessions, certain mechanisms, like device login identification, should be set up, so that after data collection and analysis methods are developed, you can only operate on the basis of the guidelines.

- **Cost-Cutting**

Businesses will be able to see how money is spent and how wasteful decisions may be avoided by examining cost centers using estimations. Even in highly regulated industries like hospitals, public sector expenditures, or compliance, AI (Ghotbi et. al. 2022) estimates will detect duplicates and propose lowering wasteful spending. A thorough analysis of agreements might help you find better opportunities.

- **Work is Flowing in New Directions**

Some employees may include the use of data - driven decision as a new approach to deal with their employment, yet it is typically more equitable, faster, and less expensive. A business that is based on rather than management hunches, big data has a better chance of growing quickly. Accepting the current working style might be tough, but it can be made easier if the Association's culture encourages self-evaluation by providing personal record records instead of an assessment meeting.

QUESTIONS ABOUT THE IMPLEMENTATION OF BIG DATA

The biggest problem with big data right now is a scarcity of expertise. Approximately 7,000 computer tech scientists and computer engineers, as well as 9,000 support employees including such data analysis associates and comparable openings, are now available.

Confidentiality, external identity security, and the originator's approval rights are also important. These may not be hazards, but rather difficulties that can be overcome, regardless of how long it takes for suitable anonymization techniques and traceable data protection to be implemented.

BIG DATA: INNOVATION, COMPETITION, AND PRODUCTIVITY'S NEXT FRONTIER

As long as the correct rules and facilitators are in place, big data will be the primary strategic premise that drives future development in efficiency, output, and consumer surplus.

Table 1. Technological Innovations of Big Data

Big data	Features	Date
Data Architecture	The feeding, cleaning, and analytics of data which is too big or complex for conventional database systems may be managed via big data frameworks.	Early 1990s
Data Governance	Data availability, usefulness, integrity, and security of data utilized in an organization are processes and factors that are managed by data governance. It covers every step from storing the data to protecting it from any accidents. It involves more than simply technology accountable for both the technology and the specific dataset. Additionally, it is used in mature organizations to guarantee the management and security of crucial data. This improves the information's clarity, which aids in outlining the decision-making procedures using data. It is a lengthy, planned procedure. For firms in the finance and insurance sectors, particularly those that must adhere to regulations, it is crucial. To oversee data throughout its life cycle, these companies are needed to establish proper data management practices. Data governance may also make it possible to authorize certain users based on classified data.	2005
Master Data Management	The requirement for organizations to facilitate the stability and excellence of their primary datasets, like commodity, resources, clients, etc., gave rise to master data management (MDM).	1980s
Metadata Management	The practice of controlling the metadata regarding data is known as metadata management. It defines and provides meaning to the information assets in your company. Your data's value is unlocked by metadata, which makes it easier to use and locate. The context that is necessary to comprehend and manage your company, data, and systems is provided by metadata. It is simpler to identify and utilize data and to offer the crucial data context that your business and IT teams need when utilizing metadata management.	1986
Processing Infrastructure	Data Processing Infrastructure includes all VFS-owned, managed, operating systems, **cloud storages,** networks, workstations, **servers,** laptops, web applications, **mobile devices,** mobile apps and websites.	1999
Data Analytics	Analyzing data collections to find patterns and take decisions about the information they have is called data analytics. Data analytics is achieved with the utilization of expert hardware and software.	1962
Data Quality	Data completeness, precision, stability, dependability, etc are all utilized to find the superiority of the data. Business firms may find data inconsistencies which require to be established and find if the data in their IT infrastructures is appropriate for the use by calculating the data superiority levels.	1972
Data Integration	Bringing together data from several sources to provide people a single perspective is known as data integration. Making data more readily accessible, simpler to consume, and easier to use by systems and people is the foundation of data integration.	1991

Data measurements have increased as a result of research by MGI and McKinsey's Enterprise Technology Office, and processing enormous quantities of ostensibly big data has become a critical prelude to innovation, which underpins new waves of profitable growth, production, and consumer surplus. Massive numbers can be lived with by leaders in all businesses, not just a few innovation managers. The volume and quality of information obtained by organizations, as well as digital advancements, online networks, and the Internet of Things (Shahab et. al., 2022), will fuel accelerated knowledge creation in the future.

MGI investigated big data in hospitals in the United States, the public sector in Europe, retail in the United States, and industrial and personal field data all over the globe. Value may be created from large data. For example, a retailer that fully utilizes big data could see a 60 percent increase in operating output. Big data has enormous promise in the public sector as well. If US health care were to use big data properly and efficiently to increase productivity and safety, it might save more than $300 billion each year. Our health spending will reduce this by approximately 8%. Government authorities in Europe's mature economies might save approximately € 100 billion by leveraging Big Data to reduce misrepresentations and errors, as well as increase the collection of assessment taxes, alone, in enhancing operational performance. Consumers of personal area data-enabled networks (Choudhary et. al., 2022) will profit by more than $600 billion. The analysis provides access to seven key pieces of information.

Data has been collected in every industry and firm, and it is a crucial growth component, alongside labor and money. By 2009, we expect practically every sector of the US economy to have 200 terabytes of stored data per firm with over 1000 workers (twice the size of Walmart's data warehouse in 1999).

There are five typical methods for generating value from big data. Big data will initially give up a great deal of value by making information more accessible. Second, as organizations develop and maintain transactional records more consistently and in depth on all types of product inventories and off-date, they will be able to catch and expose volatility, hence increasing efficiency (Gupta et. al., 2022). Leading companies use data gathering and analysis to carefully monitored testing in order to make better management choices; others employ data for low-frequency estimate at high frequency, and are actively adjusting market levers without free time. Third, Big Data allows for more precise customer classification (Bhattacharya et. al., 2021), resulting in more accurate personalized products and services. Fourth, advanced analysis will significantly improve decision-making. Big data will finally be leveraged to help create new products and services for the future generation. Manufacturers, for example, utilize data obtained from sensors (Sinha et. al., 2021) in goods to develop innovative after-sales solutions like maintenance work (preventative action taken prior to or even detected a failure).

Individual firms' success and expansion are predicated on the utilization of big data. Big Data must be treated seriously by both enterprises in terms of competitiveness and potential market gain. Electronic competitors and newcomers will employ data-driven strategies to innovate, perform, and acquire value from precise and up-to-date data in a variety of industries. In actuality, we identified early instances of such data exploitation in every field we investigated.

The utilization of big data may allow new waves of productivity growth and market excess. For example, we predict that merchants that use huge data in its totality will be able to increase their operational plans by more than 60%. Customers, as well as businesses and organizations, benefit greatly from big data. Customers may capture an economic surplus of $600 billion through personal data suppliers, for example.

Because big data applications are cross-sectored, several disciplines are expected to be included. We compared honest U.S. company competitiveness to those sectors' capacity to gather Big Data (to use an Algorithm incorporating several quantitative indicators) performance and discovered that opportunities and challenges vary by business sector. Finance and insurance, as well as the digital and electrical products and information sectors, are ready for substantial use of big data.

There'd be a lack of capabilities for organizations to utilize big data (Bansal et. al., 2022). Only between 140,000 and 190,000 people with a strong sense of reasoning will be available in the United States by 2018, while 1,5 million executives and analysts will employ big data analysis for effective decision-making (K.aushik et. al., 2021) objectives.

To fully use the potential of big data, a number of difficulties must be overcome. In the big data context, security, safety, intellectual property, and even risk tactics should be examined. Instead of just producing the necessary skills and technology (Pramanik et. al., 2022), assemblies must now arrange work procedures and benefits to enhance the utilization of big data. Access to data is critical to any organization, and resources must be allocated to make it possible to incorporate data from a variety of sources, including outsiders.

CONCLUSION

In terms of data and analysis, the industrial sector has progressed far enough, but not far enough. "There is no potential for even a large number of managers to see and know what Big Data can accomplish and revolutionize the way they work", MIT Professor Eric Byrnjolfsson wrote only this month. Manufacturers that lack a long-term view will be significantly disadvantaged in their current business environment. Instead of returning highly educated individuals to rival markets, you believe it is more acceptable to keep more roles offshore. Basic information would be buried, making it impossible to enhance or grow productivity without the presence of equipment or staff. They are seen as less reliable development network partners. The parade of actual horrors starts - by the end of the day; they will have left it behind rivalry for a long time. Due to the expansion of informal network sites, search engines, online distribution platforms, financial markets, news channels, and other information sites, data is being computed rapidly throughout the globe. Big Data is the newest frontier in science and market research. Big data analysis is critical for organized knowledge discovery in the context of sometimes occurring patterns and hidden rules. Big Data Analysis assists firms in making strategic decisions, predicting and recognizing trends, and identifying prospective business opportunities. In this article, we look at issues and challenges with big data mining as well as big data analysis tools like Map Reduce over Hadoop and HDFS, that help businesses better understand their customers and markets and make better choices. We do, however, provide various large-scale data mining approaches as well as advice on how to extract useful information from huge data. This would aid researchers in selecting the appropriate digging instruments for their projects. We handled the many possibilities that'd result in smart apps employing the framework to increase their operations and efficiency by using every accessible date as a consequence of this. We also addressed a slew of concerns and identified a number of roadblocks to the expansion of big data applications. We suggested that the generic standards for Big Data intelligent city apps be revised to concentrate on the discussion. It is necessary to develop and execute effective and dependable systems. Furthermore, these criteria often seek to resolve concerns by recommending other approaches to conquering some of the challenges and achieving better results. Finally, we discussed some of the basic, transparent issues that need to be further investigated and addressed in order to have a better knowledge of intelligent towns.

REFERENCES

Argent, R., Hetherington-Rauth, M., Stang, J., Tarp, J., Ortega, F. B., Molina-Garcia, P., Schumann, M., Bloch, W., Cheng, S., Grøntved, A., Brønd, J. C., Ekelund, U., Sardinha, L. B., & Caulfield, B. (2022). Recommendations for Determining the Validity of Consumer Wearables and Smartphones for the Estimation of Energy Expenditure: Expert Statement and Checklist of the INTERLIVE Network. *Sports Medicine (Auckland, N.Z.)*, *52*(8), 1817–1832. doi:10.100740279-022-01665-4 PMID:35260991

Bandyopadhyay, S., Goyal, V., & Dutta, S., Pramanik, S. ,& Sherazi, H. H. R. (2021). Unseen to Seen by Digital Steganography. In S. Pramanik, M. M. Ghonge, R. Ravi and K. Cengiz, (Eds.), Multidisciplinary Approach to Modern Digital Steganography, pp. 1-28. IGI Global. doi:10.4018/978-1-7998-7160-6.ch001

Bansal, R., Jenipher, B., Nisha, V., & Makhan, R. Kumbhkar, Pramanik, S., Roy, S. & Gupta, A. (2022). Big Data Architecture for Network Security. In Cyber Security and Network Security. Wiley.

Bhattacharya, A., Ghosal, A., Obaid, A. J., Krit, S., Shukla, V. K., Mandal, K., & Pramanik, S. (2021). Unsupervised Summarization Approach with Computational Statistics of Microblog Data. In D. Samanta, R. R. Althar, S. Pramanik, & S. Dutta (Eds.), *Methodologies and Applications of Computational Statistics for Machine Learning* (pp. 23–37). IGI Global. doi:10.4018/978-1-7998-7701-1.ch002

Choudhary, S., Narayan, V., Faiz, M., & Pramanik, S. (2022). Fuzzy Approach-Based Stable Energy-Efficient AODV Routing Protocol in Mobile Ad hoc Networks. In M. M. Ghonge, S. Pramanik, & A. D. Potgantwar (Eds.), *Software Defined Networking for Ad Hoc Networks*. Springer. doi:10.1007/978-3-030-91149-2_6

Dushyant, K., Muskan, G., Gupta, A., & Pramanik, S. (2022). Utilizing Machine Learning and Deep Learning in Cyber security: An Innovative Approach. In M. M. Ghonge, S. Pramanik, R. Mangrulkar, & D. N. Le (Eds.), *Cyber security and Digital Forensics*. Wiley., doi:10.1002/9781119795667.ch12

Ghotbi, N., Ho, M. T., & Mantello, P. (2022). Attitude of college students towards ethical issues of artificial intelligence in an international university in Japan. *AI & Society*, *37*(1), 283–290. doi:10.100700146-021-01168-2

Goh, G. L., Dikshit, V., Koneru, R., Peh, Z. K., Lu, W., Goh, G. D., & Yeong, W. Y. (2022). Fabrication of design-optimized multifunctional safety cage with conformal circuits for drone using hybrid 3D printing technology. *International Journal of Advanced Manufacturing Technology*, *120*(3-4), 2573–2586. doi:10.100700170-022-08831-y

Gupta, A., Verma, A., & Pramanik, S. (2022). Security Aspects in Advanced Image Processing Techniques for COVID-19. In S. Pramanik, A. Sharma, S. Bhatia, & D. N. Le (Eds.), *An Interdisciplinary Approach to Modern Network Security*. CRC Press.

Kanan, T., Mughaid, A., Al-Shalabi, R., Al-Ayyoub, M., Elbes, M., & Sadaqa, O. (2022). Business intelligence using deep learning techniques for social media contents. *Cluster Computing*. Advance online publication. doi:10.100710586-022-03626-y

K.aushik, D., Garg, V., Annu, Gupta, A. & Pramanik, S. (2021). Application of Machine Learning and Deep Learning in Cyber security: An Innovative Approach. In M. Ghonge, S. Pramanik, R. Mangrulkar and D. N. Le, (Eds.), *Cybersecurity and Digital Forensics: Challenges and Future Trends*. Wiley.

Koren, Y., Rendle, S., & Bell, R. (2022). Advances in Collaborative Filtering. In F. Ricci, L. Rokach, & B. Shapira (Eds.), *Recommender Systems Handbook*. Springer. doi:10.1007/978-1-0716-2197-4_3

Madanian, S., Nguyen, H. H., & Mirza, F. (2021). Wearable Technology. In D. Gu & M. E. Dupre (Eds.), *Encyclopedia of Gerontology and Population Aging*. Springer. doi:10.1007/978-3-030-22009-9_459

Nyangaresi, V. O. (2022). A Formally Validated Authentication Algorithm for Secure Message Forwarding in Smart Home Networks. *SN. Computer Science*, *3*, 364. doi:10.100742979-022-01269-9

Pramanik, S. (2022). An Effective Secured Privacy-Protecting Data Aggregation Method in IoT. In M. O. Odhiambo, W. Mwashita, & I. G. I. Global (Eds.), *Achieving Full Realization and Mitigating the Challenges of the Internet of Things*., doi:10.4018/978-1-7998-9312-7.ch008

Pramanik, S., & Bandyopadhyay, S. (2022). Analysis of Big Data. In I. G. I. John Wang (Ed.), *Encyclopedia of Data Science and Machine Learning*. Global.

Pramanik, S., Joardar, S., Jena, O. P., & Obaid, A. J. (2022). An Analysis of the Operations and Confrontations of Using Green IT in Sustainable Farming. *2nd International Conference on Modern Applications of Information and Communication Technology*, Al-Kadhum, Iraq.

Schloemer, T. H. (2022). Managing big data. *Nature Energy*, *7*(2), 122–123. doi:10.103841560-021-00972-w

Shahab, S., Agarwal, P., Mufti, T., & Obaid, A. J. (2022). SIoT (Social Internet of Things): A Review. In S. Fong, N. Dey, & A. Joshi (Eds.), *ICT Analysis and Applications. Lecture Notes in Networks and Systems* (Vol. 314). Springer. doi:10.1007/978-981-16-5655-2_28

Sinha, M., Chacko, E., Makhija, P., & Pramanik, S. (2021). Energy Efficient Smart Cities with Green IoT. In C. Chakrabarty (Ed.), *Green Technological Innovation for Sustainable Smart Societies: Post Pandemic Era*. Springer. doi:10.1007/978-3-030-73295-0_16

Welc, J. (2022). Financial Statement Analysis. In *Evaluating Corporate Financial Performance*. Palgrave Macmillan. doi:10.1007/978-3-030-97582-1_3

Zamani, E. D., Griva, A., & Conboy, K. (2022). (2022). Using Business Analytics for SME Business Model Transformation under Pandemic Time Pressure. *Information Systems Frontiers*. Advance online publication. doi:10.100710796-022-10255-8 PMID:35250366

Chapter 14
Towards an Effective Integration of ESP in Healthcare ICTs:
Challenges and Strategies - The Case of Algeria

Kherraz Mokhtaria
Dr. Moulay Tahar University, Algeria

ABSTRACT

Healthcare, which is considered as one of the most strategic fields, has noticed a remarkable and continuous progress over time. Yet, this sensitive field though the crucial role it plays in providing both individuals and communities with cure and care, still needs more attention from policy makers, researchers, and doctors due to the rapid and intensive progress information communication technologies (henceforth ICT) improve day after day. Hence, the present investigation attempts to highlight the use of ICT technologies in developing healthcare in Algeria as a case study and seeks to suggest better ways of using technological advancement to overcome the challenges that still hinder against its progress. The study in hand is divided into several parts. First, it tackles the recent situation of healthcare in matter of using digital technologies in accessing, processing, and storing information. Then, it tries to shed light on the different drawbacks of the misuse of ICT in healthcare settings and state the causes and consequences.

DOI: 10.4018/978-1-6684-5656-9.ch014

INTRODUCTION

Since ever, healthcare has been as a field of interest and importance. Similarly to the existence of the human being, healthcare has always gone through different phases of progress. From the use of herbal medicine in which various herbs had been used in curing sick people, and which lasted for centuries, to the use of medications after scientific research had spread throughout all fields in general and healthcare in particular. Yet, there has always been a crucial need to finding a better way in performing delicate surgeries mainly for the serious illnesses, doing medical tests, and medical x-rays (Yıkılmaz, 2020).

Besides, after the appearance of information communication technologies (henceforth ICT) healthcare has noticed a remarkable shift towards great progress when different technologies have been used in different medical centers, hospitals, etc. Indeed, the use of technologies in healthcare has improved good outcomes but there are still some difficulties that hinder against attaining specialists; here, the doctors and researchers' desires and targets in Algeria are a case study. In this vein, the following research questions have been stated:

- How is the current state of healthcare in Algeria?
- What types of technologies are used in healthcare in Algeria?
- How can the Human Resources in healthcare be trained to use the new technologies?

In accordance to the previous research questions, the following hypotheses have been suggested:

- Algeria has adopted the new Information Communication Technologies in different fields, as healthcare is considered as a strategic field hence the situation may be adequate to the demands.
- Indeed, there are noticeable interchanges between Algeria and various developed countries in matter of new technologies. Hence all sorts of technologies should be adopted in healthcare field
- New ICTs means new ways of use and new methods therefore training human resources in healthcare is crucial so as to up-date their knowledge, competencies and profeciency.

HEALTHCARE AND THE USE OF ICT

Accordingly, healthcare, particularly in Algeria, has increasingly improved good satisfaction for patients who suffer from different illnesses such as cancer, hypertension, respiratory illnesses, allergies … etc. It is worth noticeable that the implementation of ICTs in such a field has increased the effectiveness of the new adopted ways in curing sick people (Yıkılmaz, 2020).

Interpretations of Information Communication Technologies (ICTS)

Information Communication Technologies which is acronymed ICT refers to the use of computing devices such as the computer, the input, output and storage devices for specific purposes. In this sense, Hew & Brush (2017:225) state that ICT is used to refer to 'computing devices such as desktop computers, laptops, software or internet for instructional purposes'.

From another perspective and in a broader view, Tinio Victoria argues that ICT represents all types of 'technologies as radio, television, DVD, telephone (both fixed and mobile), satellite system, computer and network, hardware and software, as well as the equipment and services associated with these technologies such as video conferencing and electronic mail' (Tinion, Victoria L, 2002). Hence, Information Communications Technologies goes beyond PCs to all what is related to technology such as the radio, tv set and telephone and many other devices. Therefore, ICT means the use of different tools in order to facilitate communication processing and transmission of information via electronic means (Mayes, 2020).

Indeed, the use of information communication technologies has noticed a continuous progress since its appearance which has facilitated its spread throughout all fields. The fact that the need to highly qualified workers has increased and has become the most important criterion in recruitment.

Therefore, ICT used in the occupationnel setting helps enhence their technical abilities, performance, and professional skills, including English skills since it dominates the fields of information and communication technology. On the other hand, ICT stands as an umbrella of two components as Sakar (2012:30-31) states:

ICT s can be divided into two components: information and communication infrasructure (ICI) and information technology (IT). The former refers to physical telecommunication system and network (cellular, voice, mail, radio, and television) while the latter refers to hardware and software of information, collection, storage, processing, and presentation

ICTs IN THE REALM OF HEALTH SECTOR IN ALGERIA

Indeed, Algeria has stepped towards the digitalization of health sector during the few last decades and good results could be proved to some extent. Hence, implementing new digital technologies in health sector could be recognised through saving time in space, human resources and workplace materials. Moreover, rapid access, process and storage of huge data has transformed the field of health by providing it with technological progress and advances (Hyland, 2002).

In order to increase its benefits, Algeria has invested in the new technologies since 2005 as it was said at the World Summit on the Information Society (WSIS) where Algeria was a stakeholder. Later in 2016, Algeria was ranked the third most dynamic country in the world (according to the International Telecommunication Union (ITU)).

Since then, similar to hospitals the number of university hospitals have increased in Algeria. Furthermore, healthcare expenses have been reduced by reducing operations costs for patients and facilitating the appointments which were previously difficult to make.

In short, the use of both ICTs and digital technologies have helped all fields in general and health care in particular to be more developed and well managed.

Use of Digital Technology by Professionals in Healthcare

According to a study conducted by Pr Mohamed Amine Brahami (2022) about digitization of healthcare in Algeria more exactly in Oran, more than 67% of healthcare professional use digital equipment in their work such as desktop, laptop, tablet PC, smartphones, scanners, printers, and network equipment (routers and wifi) (Stefaniak, 2019). Indeed, the investigation came with reliable results among which are the following points:

- 75% of the healthcare professionals are satisfied with the use of various digital devices.
- 71% of the respondents confirmed that the use ICT equipments has added value to their work.
- 34% of the respondents were able to use ICTs equipments appropriately because they had undergone trainings at their establishments (haf a day or full day).

To sum up, it is clear from the results shown above that professional in healthcare in Algeria are aware of the importance the new technologies play in helping and facilitating their work yet the lack of training hinders against their needs and wants in using digital devices at the workplace.

IMPACTS OF ICTs USE ON HEALTHCARE

Actually, it is worth noticing that the implementation of different ICTs in health sector have had direct effects on the three basic elements in this field: patients, practitioners and decision makers in a sort that each element could get benefits from using such digital technologies. The patients find it easy to make appointments which has resulted in an increase in mobiles use in addition to a benefit from the medical services in short time. The practitioners, namely doctors, can ensure good results through the use of various technological devices such as x-rays and MRI where as stakholders who are responsible of implementing the new technologies in the health sector may create other reforms that may empower this sector more than before.

In the light of what was said, the Use of ICTs may have positive impacts on health care some of which may be as follow:

- Facilitate remote consultation, making diagnosis and treatments through the use of tele- medicine
- Encourage collaboration and cooperation between health workers through sharing of learning and training approaches
- Encourage more effective health research
- Facilitate the dissemination and access to different research findings
- Improve the efficiency of administrative systems in health crae through implementing new technologies

SUGGESTIONS FOR GOOD USE OF ICTs IN HEALTHCARE

Among the solutions that are required for a successful use of ICTs and digital devices within health sector is training the human resources who are supposed to use the mat their workplace. Then, what do we mean by training at the workplace?

Training Human Resources at The Workplace

Training employees at the workplace, particularly in healthcare, is an important factor that influences employees' skills, abilities, and attitudes. It also helps up-date their professional performance with the advance of technology and enables them to manage the functions of the newly technical equipment. Through training, employees can increase their performance and be able to acquire new information and skills as Armstrong (2006) states *"Performance is the matter of what people achieve and how they achieve it. Performance is, thus, the affective use of the required knowledge, skills and competencies" (Armstrong, 2006:497-498)* (Hew 2007).

In other words, having high potential employees does not mean that they will succeed. Yet, they must know what they have to do and how to do it. That is why training experts often use the phrase "workplace learning and performance". Therefore, taking the employees' performance as target to develop before their workplace training takes place would certainly facilitate the path for the trainer to point out the objectives to be attained such as developing employees' language skills, proficiency in using ICTs may lead to successful workplace training (Hew, 2007). The next figure illustrates more the requirements of good workplace training.

Figure 1. The Relation between Training Employees and their Performance

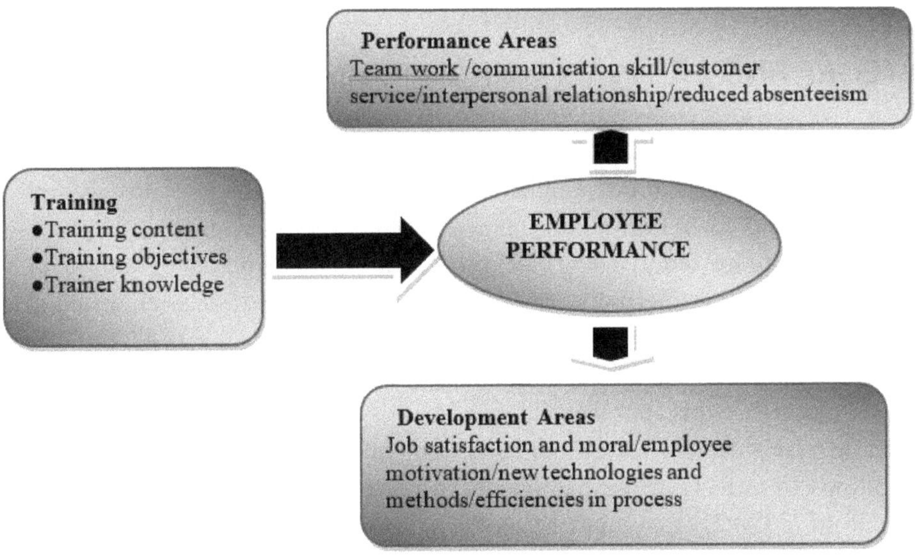

Boxall and Purcell (2016)

Besides, both practitioners and researchers in the field of work place training agree that there are a set of factors that affect the effectiveness of workers training in their professional environment, among which motivation, proficiency, professional skills and competencies should be taken into consideration (Duff, 2008).

Besides, in order to ensure good feed -back from the trainees, training target should meet their needs, wants and interests. As the training basic target is to enable workers to use ICTs appropriately then the themes of the training ought to be related to the field of ICTs. In this vein, the following themes may be of interest:

- Internet
- Teleconferencing tools
- Mobile Learning
- Electronic Learning

In short, research has proved that most workplace adult learners are motivated intrinsically by their career development, and they are extrinsically motivated by financial benefits and promotions. Besides, they have positive attitudes towards new experiences in using new technologies and they demonstrate their effectiveness in materials use.

INTEGRATING ESP IN TRAINING AT THE WORKPLACE

After its appearance, English for specific purposes (henceforth ESP) has invaded the land of teaching/learning English (EAP) and even training at the workplace (EOP) as an approach. Indeed, this approach has traced a new path within because it is basically addressed to a new category of learners with new programs and new objectives. Different interpretations from various scholars have been stated since ESP was of a great interest.

Accordingly, ESP has attracted many scholars who have already recognised the crucial importance this approach has gained for it has been more and more required by learners of different levels and categories all over the globe. Then, different definitions from distinct perspectives have been stated. In this vein, Macky and Mountford (1978:2) state that ESP is *"generally used to refer to the teaching of English for a clearly utilitarian purpose"*. On his side, Harmer (2001) argues that ESP is *"…situations where the student has some specific reasons to learn a language,* "Hutchinson & Waters (1987:19) claim that ESP is basically *"Why do these learners need to learn English?"*. Dickinson (1991:91) confirms that ESP refers to the *« skills which a learner perceives as being relevant to him »* whereas Dudley Evans & St John (1998:126) state that to teach ESP requires to *"Know learners as people, as language users & as language learners. Know how language learning and skills learning may be maximised. Know the target situation and learning environment."* Thus, from the previous quotes, it is clear that ESP is learner-centered, since the learner represents the core element in all the phases of ESP teaching.

ESP Brands

Actually, there have been many attempts to subdivide ESP into sub-branches. Yet, most of practitioners have agreed on that ESP has two main branches; English for academic purposes (EAP) and English for occupational purposes (EOP). The former is further divided into English for science and technology (EST), English for medical purposes (EMP), English for law and English for business and economy (EBE). Whereas, the latter is further divided into two main branches; English for professional purposes (EPP) and English for vocational purposes (EVP).

In this sense, Dudley-Evans and St John (1998) argue that *'ESP has traditionally been divided into two main areas: English for Academic Purposes (EAP) and English for Occupational Purposes (EOP)."* From another perspective, Cummin (2008) has differentiated between EAP and EOP in that EAP aims to provide academic proficiency, whereas, EOP leads to interpersonal skills and communication with friends and co-workers in everyday life.

English For Academic Purposes (EAP)

In fact, English for Academic Purposes (EAP) is a sub-branch within ESP. It is characterized by its focus on academic contexts that are encountered in a university setting. It is then a branch of (ELT) and its program is an integrated skills program planned for upper-intermediate to advanced English students who intend to develop their academic reading, writing, speaking and listening skills. Robinson (1988:4) points out; *"EAP is generally taught within educational institutions to students reading English in their studies.* "EAP generally occurs in the contexts of academic studies (Dessler, 2013). Besides, Robinson (1991: 2-5) describes EAP in matters of:

- EAP learners have usually current higher education level.
- EAP learners need to learn English to succeed in their academic career.
- EAP involves an attitude to learning and teaching specific language and practices required in particular academic settings (Dessler, 2013).
- Most EAP learners are undertaking fixed terms courses to prepare a particular task such as writing essays and dissertation, taking part in group discussions and oral presentations in conferences.

On their side, Hyland and Hamp-Lyons (2002) add that *"EAP is the linguistics, sociolinguistic and psycholinguistic description of English as it occurs in the contexts of academic study and scholarly exchange itself."*

To sum up, EAP is in short the theoretical part which is needed to understand or to write academic papers, whereas EAP course is more focused on theory, methodology, concepts and lexicons.

English For Occupational Purposes (EOP)

English for occupational purposes (EOP) is another sub-discipline within ESP, but it differs from (EAP) in that it is more oriented towards the application of the theoretical knowledge in the real context. Kennedy and Bolitho (1984:4) state; *"EOP is taught in a situation in which learners need to use English as part of their work profession."* In addition, EOP courses can be considered as a type of English for specific purposes (ESP) as Robinson (1991: 3) states *"... for a classification of all the main areas under ESP."* On their side, Dudley-Evans et al (1998:7) add:

"The term EOP refers to English that is not for academic purposes, it includes professional purposes in administration, medicine, law and business and vocational purposes for non-professionals in work or pre-study work"

Thus, EOP is concerned with teaching English in occupational settings, in other words, EOP learners require specific English skills in their field of study or their workplace. Teaching EOP in the computer science field for instance seeks to produce skilled manpower like good advanced-technicians, skilled managers, competent computer programmers and skillful soft-ware and hard-ware engineers.

EAP VS EOP

In the 1980's, both of English for academic purposes (EAP) and English for occupational purposes (EOP) represented sub-disciplines within English for specific purposes (ESP) and were grouped together when Carver (1983) stated three types of ESP: a) English as a restricted language b) English for academic and occupational purposes c) English with specific topics.

As a support to Carver's view, Hutchinson and Waters (1987:16) state that there is not a clear cut-distinction between EAP and EOP; they state that *"People can work and study simultaneously; it is also likely that in many cases the language learnt for immediate use in a study environment will be used later when the student takes up or returns to a job."*

According to Hutchinson and Waters, there is a complementarity between EAP and EOP, they may even go hand in hand if the learner studies and works at the same time or even if this learner intends to use English at the workplace after having acquired it in an academic setting like the university.

Hyland and Hamp-Lyons state that EAP differs from EOP in its focus on academic contexts, they add; *"EAP is the linguistic, sociolinguistic and psycholinguistic description of English as it occurs in the contexts of academic study and scholarly exchange itself."* Hamp Lyons on his side assures that the English Language skills of EAP learners will soon develop as the demand for EAP is getting higher by the non-native English-speaking academics.

In short, EAP learners require the kind of English to use in academic settings while EOP learners require the kind of English they need in their work profession. Actually, the present study focuses more on EOP for the target context, which will be studied, is the use of English in the workplace training. Besides, Strevens (1977) has confirmed Hutchinson and Waters' view through dividing ESP into EAP and EOP and further dividing EAP into three sub-divisions: pre-study, in-study and post-study besides EOP is split into pre-experience, simultaneous and post experience. For the case of integrating ESP in training human resources within health sector in Algeria, the next model is suggested to be adopted.

ADDIE As a Workplace Training Model

ADDIE is an acronym that stands for five concepts:

- Analysis
- Design
- Development
- Implementation
- Evaluation

Indeed, ADDIE refers to the five stages that make up the training process. From the name itself, it can be understood that each stage is classified in that clear order so that it cannot change position. That is to mean, if we refer to stage three in ADDIE, we then talk about development and we do not talk about any other stage. Besides, these five stages make a circular pattern so that when applying them in training employees, they may be repeated several times till the employee reaches his target. Hence, when the required feedback is attained, then the final target of the training process is reached. The next figure shows how ADDIE model is shaped.

From the figure above, it is clear that the first four stages which are analysis, design, development and implementation are followed by each other whereas the fifth stage which is evaluation in located in the center and has a relation with all the other stages this means that each stage can be evaluated on its own and this is called formative evaluation. Besides, there is another evaluation that is done by the

Figure 2. The Five Stages of ADDIE Model

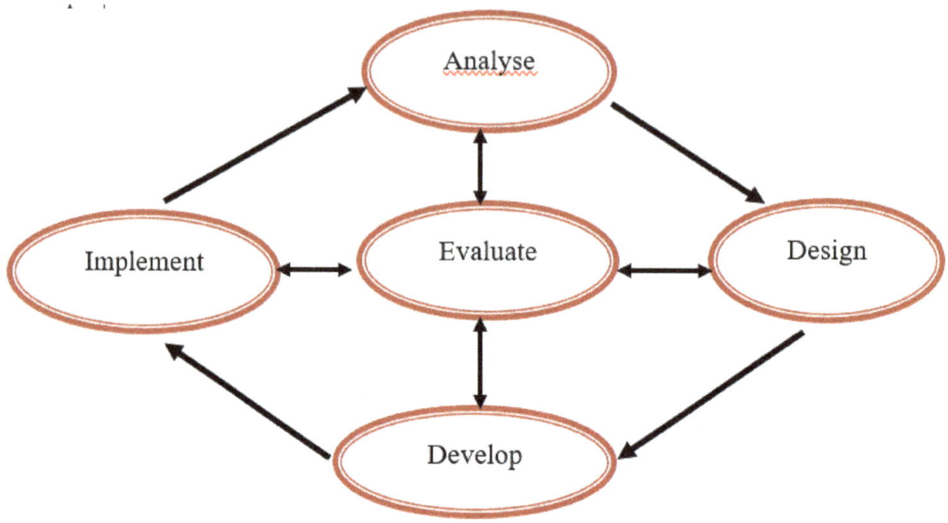

end of the training that is to mean when the four phases are done, this evaluation is called summative evaluation. Yet, what each stage mean?

- **Analysis** is the first stage that trainers/designers step towards creating a training program (Hutchinson, 1987). In this phase, it is very important to put some questions as background in order to launch the model in the right way, the questions are:

  Why is training needed?
  Who is supposed to be trained?
  What should be the content of the training program?
  Where will the training take place?
  When will the training take place?
  How will the training be processed?

Therefore, asking such questions may facilitate gathering the needed content in order to set up the suitable needs analysis that would be conducted with the employees as a first step.

- **Design**, which is supposed to be the second step, means to be aware about the reason for which the workplace training is designed. In other words, it helps create the blue print in order to build the courses outlines.

- **Development** refers to the phase in which the training is developed out either through an e-learning or a face to face learning or a video learning course. Hence, using these three ways of training, employees would learn different organizational processes, systems, policies and skills (Leisink, 2021).
- **Implementation** is the practical phase, it refers to the performance of the training towards the target audience which is the employees (Leisink, 2021).
- **Evaluation,** which means the assessment of all what has preceded, helps know whether it works or not. It does not evaluate the employees only but it also assesses the training courses as well. In addition to the evaluation, there is also the matrics matter which shows the period the training has lasted, what the employees have acquired during the training period, and whether the training which has been offered to the employees is useful and up to their jobs or not (Leisink, 2021).

From another perspective and in order to better explain the ADDIE training, we need to know what ADDIE is not. Indeed, ADDIE is not a process, it is rather a basic work that can be processed in many different ways, but it depends on the field it is used for. In fact, there is process for every single project. As for health sector, it needs a specific process different from other processes but has the same common background which is ADDIE. Moreover, the choice of ADDIE model has not been randomly done. Indeed, there are a number of factors, stated by scholars, and that helped the researcher in her choice:

According to Culatta and Kearsley (2019a), instructional design models are considered as guidelines or frameworks adopted by organizations in order to make up the process of creating the training programs and activities as practice for the benefit of the employees. In fact, the ADDIE model is seen as the best known compared with all the other models (Culatta, 2018a)

Besides, Gagne, Wager, Golas & Keller (2005) argue that ADDIE is an iterative or circular process that is advantageous because of its flexibility. In other words, this model may change according to the needs of each instructional design. They add that ADDIE may be used as a chain where the end of each phase becomes the beginning of the next phase which means that it can be used in sequences (moving forward or backward). As far as the evaluation phase is concerned, ADDIE comprises two types of evaluations: Formative assessment and Summative assessment.

Workplace English Training Phases

Similar to ADDIE model, Dessler (2013) has divided the training process into five phases:

- Phase one: Analyze the training needs. Indeed, training needs are usually collected in systematic process between the trainer and the stakeholders of the organization.
- Phase two: Design the overall training program. In fact, a good program is the one that intend to achieve measurable targets, if the improved achievements in matter of proficiency can be measured then the selected program is good (Michael, 2009).
- Phase three: Develop the course (providing the needed materials/equipments). Health care work place training program should focus on authentic scenarios taking into consideration the selection of training program, materials, tasks and methods.
- Phase four: Implement the training by training the target sample of workers. In the case of the current study the target participants are limited in the health sector workers mainly those who are supposed to use ICTs and digital devices (Michael, 2009).
- Phase five: Evaluate the training effectiveness. This evaluation cannot be done only after a certain period that may be six months to one year. This period is enough to see the outcomes of the workplace training and measure its effectiveness.

Other fruitful suggestions may be:

- Share information and experiences with foreign experts via international on-line conferences
- Strengthen human resources awareness skills in ICTs use
- Encourage researches in the field of ICTs and digital technologies
- Assess the impact of ICT on health (continuous assessment)
- Provide hospitals, university hospitals and medical centers with different ICTs equipments.

Benefits of Training at The Workplace

Training in the workplace has a central meaning in the discussion of the role of Human Resources in any organization. Training is the key reason for success because there is a continuous change in the strategies of the management in any field and health sector is in a central position in addition to education. Workers in healthcare have a direct connection to the field's quality of service and to the field's technological investment. For that reason, stakeholders should adapt training programs for the benefits of the human resources as well as for the society in order to:

- Increase confidence and motivation.
- Decrease risks because well trained workers can use the material and equipment without waste.
- Bring a sense of security and therefore reduce labor turnover and absenteeism.
- Involve the workers in the change process and provide them with the skills needed in order to adjust with the new situations.
- Enhance their responsibility and therefore increase pay and promotion
- Improve the availability and quality of the staff.

CONCLUSION

To sum up, it is true that with the rapid progress of technology noticed day after day and which is affecting all fields all over the globe, it requires more attention and care in order to go hand in hand with any change that may appear at any time. Healthcare which is considered as a strategic field is indeed facing many challenges. Therefore, the need for new technologies and technological devices is increasing for the multiplication of diseases is nowadays demanding more and more care.

Algeria, as a case study in the current study, though it has proved remarkable progress in implementing new information communication technologies in matter of both electronic information and digital devices health sector still needs more care particularly with the emergence of new pandemics as the covid 19 pandemic which has resulted in thousands of deaths, thanks to the use digital tools which helped slow the spread of this virus otherwise a catastrophe could happen.

Therefore, to gain the respect and trust of patients there should be a serious management of ICTs use.

REFERENCES

Carver, D. (1983). Some propositions about ESP. *The ESP Journal*, *2*(2), 131–137. doi:10.1016/0272-2380(93)90003-P

Dessler, G. (2013). Human resource management (13th Ed). Pearson.

Dickinson, L. (1987). Self-instruction in language learning. *Canadian Modern Language Review*, *44*(2), 381–382.

Dudley-Evans, T., St John, M. J., & Saint John, M. J. (1998). *Developments in English for specific purposes: A multi-disciplinary approach*. Cambridge university press.

Duff, P., & Hornberger, N. (2008). Encyclopedia of language and education. Language socialization, 8, 243-255.

Harmer, J. (2001). *The practice of English language teaching* .Pearson.

Hew, K. F., & Brush, T. (2007). Integrating technology into K-12 teaching and learning: Current knowledge gaps and recommendations for future research. *Educational Technology Research and Development*, *55*(3), 223–252. doi:10.100711423-006-9022-5

Hutchinson, T., & Waters, A. (1987). *English for specific purposes*. Cambridge university press. doi:10.1017/CBO9780511733031

Hyland, K., & Hamp-Lyons, L. (2002). EAP: Issues and directions. *Journal of English for Academic Purposes*, *1*(1), 1–12. doi:10.1016/S1475-1585(02)00002-4

Leisink, P., & Boxall, P. (2021). HRM and social legitimacy in the public sector. In *Research Handbook on HRM in the Public Sector*. Edward Elgar Publishing.

Mayes, J. L. (2020). *UDL and Motivation: Student Perceptions of the Impact of Universal Design for Learning on Motivation of First-Year Community College Students in Rural East Tennessee* [Doctoral dissertation, East Tennessee State University, United States].

Michael, A. (2009). *Armstrong's handbook of human resource management practice*.

Monova-Zheleva, M., Zhelev, Y., & Nikolova, E. (2021). Fostering technology-enhanced learning and digital innovations in schools–key factors and challenges. *Mathematics and Education in Mathematics*, *50*, 215–221.

Palmer, J. D. (1978). English for Specific Purposes: A Case Study Approach.

Stefaniak, J. E. (2019). Instructional design theory. In *Open and Distance Education Theory Revisited* (pp. 85–94). Springer.

Strevens, P. (1977). Special-purpose language learning: A perspective. *Language Teaching*, *10*(3), 145–163.

Tinio, V. L. (2002). *ICT in Education. United Nations Development Programme Bureau for development policy. Lind, A. (2008). Literacy for all: Making a difference*. Unesco.

Yıkılmaz, İ. (2020). New era: The transformation from the information society to super smart society (society 5.0). In G. Mert, E. Şen, & O. Yılmaz (Eds.), *Data, Information and Knowledge Management* (pp. 85–112).

KEY TERMS AND DEFINITIONS

Competencies: Means being able to do something successfully and efficiently. For people who work, to be competent means to improve good outcomes whether in matter of job-related tasks or language master.

Hardware: Represent the physical components of the computer, it is what we can see and touch such as the CPU, the monitor.

Human Resources: Human resources are people investment within business and organization. They are concerned with all the policies and practices that include selecting suitable and required candidates, orienting new employees, planning, training, managing salaries, providing employees benefits and rewards.

Information Communication Technologies: Is the use of diverse technological tools in storing, creating, transmitting, and sharing information. The technological tools comprise computers, smart phones, etc.

Knowledge: Means the amount of information someone can know either through experience or studies about a subject matter, a language, or a job.

Performance: Performance is the matter of what employees can achieve and how they achieve it. Performance means the effective use of the needed skills, knowledge, and competencies.

Workplace Training: Is to train employees at their workplace, i.e., in their company. The program of their training is mainly designed to fulfill their needs, wants, and desires; it is generally technical.

Compilation of References

Abdurakhmonov, I. Y. (2016). *Bioinformatics: basics, development, and future*. InTech.

Abraham, A. (2005). Artificial neural networks. Handbook of measuring system design, 15.

Agrawal, J. S. (2018). an AI Powered Electronic Stethoscope. *Anaesthesia, Pain & Intensive Care*, *22*(3), 412–413.

Agrawal, T. (2021). Hyperparameter Optimization Using Scikit-Learn. In *Hyperparameter Optimization in Machine Learning*. Apress. doi:10.1007/978-1-4842-6579-6_2

Ahmad, A., Garhwal, S., Ray, S. K., Kumar, G., Malebary, S. J., & Barukab, O. M. (2021). The Number of Confirmed Cases of Covid-19 by using Machine Learning: Methods and Challenges. *Archives of Computational Methods in Engineering*, *28*(4), 2645–2653. doi:10.100711831-020-09472-8 PMID:32837183

Al-Amri, S. S., & Kalyankar, N. V. (2010). Image segmentation by using threshold techniques. *Journal of Computing, 2*(5).

Alazzam, M. B., Tayyib, N., Alshawwa, S. Z., & Ahmed, M. (2022). Nursing care systematization with case-based reasoning and artificial intelligence. *Journal of Healthcare Engineering*, *2022*, 2022. doi:10.1155/2022/1959371 PMID:35310193

Albawi, S., Mohammed, T. A., & Al-Zawi, S. (2017, August). Understanding of a convolutional neural network. In *2017 International Conference on Engineering and Technology (ICET)* (pp. 1-6). IEEE.

Alifrangis, C., McGovern, U., Freeman, A., Powles, T., & Linch, M. (2019). Molecular and histopathology directed therapy for advanced bladder cancer. *Nature Reviews. Urology*, *16*(8), 465–483. doi:10.103841585-019-0208-0 PMID:31289379

Alom, M. Z., Taha, T. M., Yakopcic, C., Westberg, S., Sidike, P., Nasrin, M. S., Hasan, M., Van Essen, B. C., Awwal, A. A. S., & Asari, V. K. (2019). A state-of-the-art survey on deep learning theory and architectures. *Electronics (Basel)*, *8*(3), 292. doi:10.3390/electronics8030292

Al-Saffar, A. A. M., Tao, H., & Talab, M. A. (2017, October). Review of deep convolution neural network in image classification. In *2017 International Conference on Radar, Antenna, Microwave, Electronics, and Telecommunications (ICRAMET)* (pp. 26-31). IEEE. 10.1109/ICRAMET.2017.8253139

Anandaraman, C., Sankar, A. V. M., & Natarajan, R. (2012). A new evolutionary algorithm based

Anderson & Goodman. (2002). *Ethics and information technology: A case-based approach to a health care system in transition.* Springer.

Anderson, J. G., & Goodman, K. W. (Eds.). (2002). *Ethics and information technology: a case-based approach to a health care system in transition.* Springer New York.

Anyanwu, M. & Shiva, S. (2009). Comparative Analysis of Serial Decision Tree Classification Algorithms. *International Journal of Computer Science and Security, (IJCSS), 3*(3).

Argent, R., Hetherington-Rauth, M., Stang, J., Tarp, J., Ortega, F. B., Molina-Garcia, P., Schumann, M., Bloch, W., Cheng, S., Grøntved, A., Brønd, J. C., Ekelund, U., Sardinha, L. B., & Caulfield, B. (2022). Recommendations for Determining the Validity of Consumer Wearables and Smartphones for the Estimation of Energy Expenditure: Expert Statement and Checklist of the INTERLIVE Network. *Sports Medicine (Auckland, N.Z.), 52*(8), 1817–1832. doi:10.100740279-022-01665-4 PMID:35260991

Askitopoulou, H., & Vgontzas, A. N. (2018). The relevance of the Hippocratic Oath to the ethical and moral values of contemporary medicine. Part I: The Hippocratic Oath from antiquity to modern times. *European Spine Journal, 27*(7), 1481–1490. doi:10.100700586-017-5348-4 PMID:29080001

Azar, A. T., & Eljamel, M. S. (2012). Medical robotics. In *Prototyping of robotic systems: applications of design and implementation* (pp. 253–287). IGI Global. doi:10.4018/978-1-4666-0176-5.ch009

Azbeg, K., Ouchetto, O., & Andaloussi, S. J. (2022). *Access Control and Privacy-Preserving Blockchain-Based System for Diseases Management.* IEEE Transactions on Computational Social Systems.

Bacanin, N., Vukobrat, N., Zivkovic, M., Bezdan, T., & Strumberger, I. (2022). Improved Harris Hawks Optimization Adapted for Artificial Neural Network Training. In C. Kahraman, S. Cebi, S. Cevik Onar, B. Oztaysi, A. C. Tolga, & I. U. Sari (Eds.), *Intelligent and Fuzzy Techniques for Emerging Conditions and Digital Transformation. INFUS 2021. Lecture Notes in Networks and Systems* (Vol. 308). Springer. doi:10.1007/978-3-030-85577-2_33

Baeta, F., Correia, J., Martins, T., & Machado, P. (2021). TensorGP – Genetic Programming Engine in TensorFlow. In P. A. Castillo & J. L. Jiménez Laredo (Eds.), Lecture Notes in Computer Science Vol. 12694. Springer. doi:10.1007/978-3-030-72699-7_48

Bai, X., Wang, Y., Song, Z., Feng, Y., Chen, Y., Zhang, D., & Feng, L. (2020). The basic properties of gold nanoparticles and their applications in tumor diagnosis and treatment. *International Journal of Molecular Sciences*, *21*(7), 2480. doi:10.3390/ijms21072480 PMID:32260051

Ball, J. E., Anderson, D. T., & Chan, C. S. Sr. (2017). Comprehensive survey of deep learning in remote sensing: Theories, tools, and challenges for the community. *Journal of Applied Remote Sensing*, *11*(4), 042609. doi:10.1117/1.JRS.11.042609

Bandyopadhyay, S., Goyal, V., & Dutta, S., Pramanik, S. ,& Sherazi, H. H. R. (2021). Unseen to Seen by Digital Steganography. In S. Pramanik, M. M. Ghonge, R. Ravi and K. Cengiz, (Eds.), Multidisciplinary Approach to Modern Digital Steganography, pp. 1-28. IGI Global. doi:10.4018/978-1-7998-7160-6.ch001

Bansal, R., Jenipher, B., Nisha, V., & Makhan, R. Kumbhkar, Pramanik, S., Roy, S. & Gupta, A. (2022). Big Data Architecture for Network Security. In Cyber Security and Network Security. Wiley.

Basha, S. S., Dubey, S. R., Pulabaigari, V., & Mukherjee, S. (2020). Impact of fully connected layers on performance of convolutional neural networks for image classification. *Neurocomputing*, *378*, 112–119. doi:10.1016/j.neucom.2019.10.008

Bassi, P. R. A. S., & Attux, R. (2022). A deep convolutional neural network for COVID-19 detection using chest X-rays. *Research on Biomedical Engineering*, *38*(1), 139–148. doi:10.100742600-021-00132-9

BBC News. (2020). *Coronavirus disease named Covid-19.* https://www.bbc.co.uk/news/world-asia-china-51466362

Bell, W. J., Roth, L. M., & Nalepa, C. A. (2007). *Cockroaches: ecology, behavior, and natural history*. JHU Press.

Benahmed, K., Merabti, M., & Haffaf, H. (2012). Inspired social spider behavior for secure wireless sensor networks. [IJMCMC]. *International Journal of Mobile Computing and Multimedia Communications*, *4*(4), 1–10.

Beni, G., & Wang, J. (1993). *Swarm intelligence in cellular robotic systems. InRobots and Biological Systems: Towards a New Bionics?* Springer Berlin Heidelberg.

Bhattacharya, A., Ghosal, A., Obaid, A. J., Krit, S., Shukla, V. K., Mandal, K., & Pramanik, S. (2021). Unsupervised Summarization Approach with Computational Statistics of Microblog Data. In D. Samanta, R. R. Althar, S. Pramanik, & S. Dutta (Eds.), *Methodologies and Applications of Computational Statistics for Machine Learning* (pp. 23–37). IGI Global. doi:10.4018/978-1-7998-7701-1.ch002

Biundo, E., Pease, A., Segers, K., & de Groote, M. (2020). The socio-economic impact of AI in healthcare. *Deloitte*.

Blackwell, T., & Bentley, P. J., Improvised music with swarms, In David B. Fogel, Mohamed A. El-Sharkawi, Xin Yao, Garry Greenwood, Hitoshi Iba, PaulMarrow, and Mark Shackleton (eds.), Proceedings of the 2002 Congress on Evolutionary Computation CEC 2002, pages 1462–1467, IEEE Press, 2002.

Bouarara, H. A., & Bouarara, Y. (2016). Swarm Intelligence Methods for Unsupervised Images Classification: Applications and Comparative Study. [IJOCI]. *International Journal of Organizational and Collective Intelligence*, *6*(2), 50–74.

Bouarara, H. A., Hamou, R. M., & Amine, A. (2014). Text Clustering using Distances Combination by Social Bees: Towards 3D Visualisation Aspect. [IJIRR]. *International Journal of Information Retrieval Research*, *4*(3), 34–53.

Bouarara, H. A., Hamou, R. M., & Amine, A. (2015). A Novel Bio-Inspired Approach for Multilingual Spam Filtering. [IJIIT]. *International Journal of Intelligent Information Technologies*, *11*(3), 45–87.

Bouarara, H. A., Hamou, R. M., & Amine, A. (2015). New Swarm Intelligence Technique of Artificial Social Cockroaches for Suspicious Person Detection Using N-Gram Pixel with Visual Result Mining. [IJSDS]. *International Journal of Strategic Decision Sciences*, *6*(3), 65–91.

Bouarara, H. A., Hamou, R. M., & Amine, A. (2015). Novel Bio-Inspired Technique of Artificial Social Cockroaches (ASC). [IJOCI]. *International Journal of Organizational and Collective Intelligence*, *5*(2), 47–79.

Bouarara, H. A., Hamou, R. M., Rahmani, A., & Amine, A. (2015). Boosting Algorithm and Meta-Heuristic Based on Genetic Algorithms for Textual Plagiarism Detection. [IJCINI]. *International Journal of Cognitive Informatics and Natural Intelligence*, *9*(4), 65–87.

Bourjot, C., Chevrier, V., & Thomas, V. (2003). A new swarm mechanism based on social spiders colonies: From web weaving to region detection. *Web Intelligence and Agent Systems: An International Journal*, *1*(1), 47–64.

Brandão, M., Guisseve, A., Bata, G., Firmino-Machado, J., Alberto, M., Ferro, J., Garcia, C., Zaqueu, C., Jamisse, A., Lorenzoni, C., Piccart-Gebhart, M., Leitão, D., Come, J., Soares, O., Gudo-Morais, A., Schmitt, F., Tulsidás, S., Carrilho, C., & Lunet, N. (2021). Survival impact and cost-effectiveness of a multidisciplinary tumor board for breast cancer in Mozambique, Sub-Saharan Africa. *The Oncologist*, *26*(6), e996–e1008. doi:10.1002/onco.13643 PMID:33325595

Breiman, L., Friedman, J., Olshen, R., & Stone, C. (1984). *Classification and Regression Trees.* Chapman and Hall.

Bremermann, H. (1974). Chemotaxis and optimization. *Journal of the Franklin Institute*, *297*(5), 397–404.

Brewer, E., Demmer, M., Du, B., Ho, M., Kam, M., Nedevschi, S., Pal, J., Patra, R., Surana, S., & Fall, K. (2005). The case for technology in developing regions. *Computer*, *38*(6), 25–38. doi:10.1109/MC.2005.204

Brown, K. C. (2018). *A Consideration of Mason's Ethical Framework: The Importance of PAPA Factors in the 21st Century: A Seven-Year Study*. University of Florida.

Cai, C. J., Winter, S., Steiner, D., Wilcox, L., & Terry, M. (2019). "Hello AI": uncovering the onboarding needs of medical practitioners for human-AI collaborative decision-making. *Proceedings of the ACM on Human-computer Interaction*, *3*(CSCW), 1-24. 10.1145/3359206

Campbell, R. G. (2019, December). Robotic surgery of the anterior skull base. *International Forum of Allergy & Rhinology*, *9*(12), 1508–1514. doi:10.1002/alr.22435 PMID:31633887

Cao, J., Chen, J., & Li, H. (2014). An Adaboost- Backpropagation Neural Network for Automated Image Sentiment Classification. *TheScientificWorldJournal*, *2014*, 1–9. doi:10.1155/2014/364649 PMID:25162047

Carvajal, O., Melin, P., Miramontes, I., & Prado-Arechiga, G. (2021). Optimal design of a general type-2 fuzzy classifier for the pulse level and its hardware implementation. *Engineering Applications of Artificial Intelligence*, *97*, 104069. doi:10.1016/j.engappai.2020.104069

Carver, D. (1983). Some propositions about ESP. *The ESP Journal*, *2*(2), 131–137. doi:10.1016/0272-2380(93)90003-P

Chandrasekhar, K., Pavan, R., Bharathi, P., & Triveni, K. V. (2019). Histopathologic Cancer Detection. *IRJCS: International Research Journal of Computer Science*, *6*, 102–124.

Chatterjee, A. (2017) Use of Artificial Intelligence to reduce Medical Errors. *Health Vectors.* https://blog.myhealthvectors.com.

Chauhan, R., Ghanshala, K. K., & Joshi, R. C. (2018, December). Convolutional neural network (CNN) for image detection and recognition. In *2018 First International Conference on Secure Cyber Computing and Communication (ICSCCC)* (pp. 278-282). IEEE. 10.1109/ICSCCC.2018.8703316

Cheng, Z., Zhang, Y., Zhou, C., Zhang, W. & Gao, S. (2010). Classification of Skin Sensitizers on the Basis of Their Effective Concentration 3 Values by Using Adaptive Boosting Method. *International Journal of Digital Content Technology and its Applications (JDCTA), 4*(2), 109 – 121.

Chew, H. S. J., & Achananuparp, P. (2022). Perceptions and Needs of Artificial Intelligence in Health Care to Increase Adoption: Scoping Review. *Journal of medical Internet research*, *24*(1), e32939. https://doi.org/10.2196/32939.

Chirita, P. A., Nejdl, W., & Zamfir, C. (2005). Preventing shilling attacks in online recommender systems. In WIDM'05 (pp. 67–74). doi:10.1145/1097047.1097061

Choi, J. Y., Kim, D. H., Plataniotis, K. N., & Ro, Y. M. (2012). *Combining Multiple Feature Representations and AdaBoost Ensemble Learning for Reducing False-Positive Detections in Computer-aided Detection of Masses on Mammograms.* 34th Annual International Conference of the IEEE EMBS. San Diego, California USA.

Choudhary, S., Narayan, V., Faiz, M., & Pramanik, S. (2022). Fuzzy Approach-Based Stable Energy-Efficient AODV Routing Protocol in Mobile Ad hoc Networks. In M. M. Ghonge, S. Pramanik, & A. D. Potgantwar (Eds.), *Software Defined Networking for Ad Hoc Networks*. Springer. doi:10.1007/978-3-030-91149-2_6

Civantos, F. J., Stoeckli, S. J., Takes, R. P., Woolgar, J. A., De Bree, R., Paleri, V., Devaney, K. O., Rinaldo, A., Silver, C. E., Mondin, V., Werner, J. A., & Ferlito, A. (2010). What is the role of sentinel lymph node biopsy in the management of oral cancer in 2010? *European Archives of Oto-Rhino-Laryngology*, *267*(6), 839–844. doi:10.100700405-010-1215-1 PMID:20204392

Cockburn, I. M., Henderson, R., & Stern, S. (2018). The impact of artificial intelligence on innovation: An exploratory analysis. In *The economics of artificial intelligence: An agenda* (pp. 115–146). University of Chicago Press.

Cohen, D. (2020). HIPAA Reform or a Patchwork Scheme: A Look at Preemption, Scope, and the Inclusion of a Private Right of Action in a New Federal Data Privacy Law. *American University Washington College of Law*.

Colorni, A., Dorigo, M., Maniezzo, V., & Trubian, M. (1994). Ant System for Job-shop Scheduling. *Belgian Journal of Operations Research, Statistics and Computer Science*, *34*(1), 39–53.

Costa, J. S., Nadia, N. N., & Mourelle, L. (2010). Automatic Adaptive Modeling of Fuzzy Systems Using Particle Swarm Optimization. *Transactions on Computational Science VIII. Lecture Notes in Computer Science*, *6260*, 71–84. doi:10.1007/978-3-642-16236-7_5

Coventry, L., & Branley, D. (2018). Cybersecurity in healthcare: A narrative review of trends, threats, and ways forward. *Maturitas*, *113*, 48–52. doi:10.1016/j.maturitas.2018.04.008 PMID:29903648

Cramer, N. L. (1985, July). A representation for the adaptive generation of simple sequential programs. In *Proceedings of the First International Conference on Genetic Algorithms* (pp. 183-187).

Cubric, M. (2020) Drivers, Barriers and social considerations for AI adoption in business and management: A tertiary study. *Technology in Society, 62.*

Cumani, A. (1991). Edge detection in multispectral images. *CVGIP. Graphical Models and Image Processing*, *53*(1), 40–51. doi:10.1016/1049-9652(91)90018-F

Dada, E. G., Bassi, J. S., Chiroma, H., Adetunmbi, A. O., & Ajibuwa, O. E. (2019). Machine learning for email spam filtering: Review, approaches and open research problems. *Heliyon*, *5*(6), e01802. doi:10.1016/j.heliyon.2019.e01802 PMID:31211254

Darwin, C. (2009). The origin of species by means of natural selection: or, the preservation of favored races in the struggle for life.

Darwin, C. (1859). *On the origin ofspecies by means of natural selection, or the preservation of favoured races in the struggle for life*. D. Appleton and Company.

Darwin, C. (1991). *On the origin of species by means of natural selection. 1859*. Murray.

Das, M., Roy, A., Maity, S., Kar, S., & Sengupta, S. (2021). Solving fuzzy dynamic ship routing and scheduling problem through new genetic algorithm. *Decision Making: Applications in Management and Engineering*. doi:. doi:10.31181/dmame181221030d

Davis, J., & Goadrich, M. (2006, June). The relationship between Precision-Recall and ROC curves. In *Proceedings of the 23rd international conference on Machine learning* (pp. 233-240). 10.1145/1143844.1143874

de Groot, N. G., & Bontrop, R. E. (2020). *COVID-19 pandemic: is a gender-defined dosage effect responsible for the high mortality rate among males?* Academic Press.

Dehariya, V. K., Shrivastava, S. K., & Jain, R. C. (2010, November). Clustering of image data set using k-means and fuzzy k-means algorithms. In *2010 International conference on computational intelligence and communication networks* (pp. 386-391). IEEE. 10.1109/CICN.2010.80

Delany, S. J., Buckley, M., & Greene, D. (2012). SMS spam filtering: Methods and data. *Expert Systems with Applications*, *39*(10), 9899–9908. doi:10.1016/j.eswa.2012.02.053

Dessler, G. (2013). Human resource management (13th Ed). Pearson.

Dhande, M., Dandekar, D., & Badjate, S. (2012). Performance Improvement of Ann Classifiers using Pso. *Proceedings of National Conference on Innovative Paradigms in Engineering & Technology*, (pp. 32-36). International Journal of Computer Applications.

Dhande, J. D., & Dandekar, D. R. (2011). PSO Based SVM as an Optimal Classifier for Classification of Radar Returns from Ionosphere. *International Journal on Emerging Technologies*, *2*(2), 1–3.

Dickinson, L. (1987). Self-instruction in language learning. *Canadian Modern Language Review*, *44*(2), 381–382.

Diehl, F. (2019). Edge contraction pooling for graph neural networks. *Cornell University ARIV.*

Dilsizian, S. E., & Siegel, E. L. (2014). Artificial intelligence in medicine and cardiac imaging: Harnessing big data and advanced computing to provide personalized medical diagnosis and treatment. *Current Cardiology Reports*, *16*(1), 1–8. doi:10.100711886-013-0441-8 PMID:24338557

Domingo-Ferrer, J., Farras, O., Ribes-González, J., & Sánchez, D. (2019). Privacy-preserving cloud computing on sensitive data: A survey of methods, products and challenges. *Computer Communications*, *140*, 38–60. doi:10.1016/j.comcom.2019.04.011

Dorigo, M. Optimization, learning and natural algorithms (in Italian), Ph.D. Thesis, Dipartimento diElettronica, Politecnico di Milano, Italy, 1992.

Dorigo, M., Maniezzo, V., & Colorni, A. Positive feedback as a search strategy, Tech. Report 91-016, Dipartimento di Elettronica, Politecnico di Milano, Italy, 1991.

Dorigo, M., & Gambardella, L. M. (1997). Ant colonies for the travelling salesman problem. *Bio Systems*, *43*(2), 73–81.

Dorigo, M., & Gambardella, L. M. (1997). Ant colony system: A cooperative learning approach to the traveling salesman problem. Evolutionary Computation. *IEEE Transactions on*, *1*(1), 53–66.

Dorigo, M., Maniezzo, V., & Colorni, A. (1996). Ant System: Optimization by a colony of cooperating agents. *IEEE Transactions on Systems, Man, and Cybernetics. Part B, Cybernetics*, *26*, 29–41.

Dorigo, M., Maniezzo, V., & Colorni, A. (1996). Ant system: optimization by a colony of cooperating agents. Systems, Man, and Cybernetics, Part B: Cybernetics. *IEEE Transactions on*, *26*(1), 29–41.

Duan, W., Zhang, L., Colman, J., Gulli, G., & Ye, X. (2021, September). Multi-modal Brain Segmentation Using Hyper-Fused Convolutional Neural Network. In *International Workshop on Machine Learning in Clinical Neuroimaging* (pp. 82-91). 10.1007/978-3-030-87586-2_9

Dudley-Evans, T., St John, M. J., & Saint John, M. J. (1998). *Developments in English for specific purposes: A multi-disciplinary approach*. Cambridge university press.

Duff, P., & Hornberger, N. (2008). Encyclopedia of language and education. Language socialization, 8, 243-255.

Dushyant, K., Muskan, G., Gupta, A., & Pramanik, S. (2022). Utilizing Machine Learning and Deep Learning in Cyber security: An Innovative Approach. In M. M. Ghonge, S. Pramanik, R. Mangrulkar, & D. N. Le (Eds.), *Cyber security and Digital Forensics*. Wiley. doi:10.1002/9781119795667.ch12

Dutta, S., Pramanik, S., & Bandyopadhyay, S. K. (2021). Prediction of Weight Gainduring COVID-19 for Avoiding Complication in Health. *International Journal of Medical Science and Current Research*, *4*(3), 1042–1052.

Eberhart, R. C., & Shi, Y. Tracking and optimizing dynamic systems with particle swarms, *Proc. Congress on Evolutionary Computation 2001*, Seoul, Korea, 2001

Ekblaw, A., Azaria, A., Halamka, J. D., & Lippman, A. (2016, August). A Case Study for Blockchain in Healthcare: "MedRec" prototype for electronic health records and medical research data. In *Proceedings of IEEE open & big data conference* (Vol. 13, p. 13).

Engelbrecht, A. P. (2007). *Computational intelligence: an introduction*. John Wiley & Sons.

Fan, L. (2019, July). Practical image obfuscation with provable privacy. In IEEE international conference on multimedia and expo (ICME) (pp. 784-789). IEEE. doi:10.1109/ICME.2019.00140

Farshi, T. R., Drake, J. H., & Özcan, E. (2020). A multimodal particle swarm optimization-based approach for image segmentation. *Expert Systems with Applications*, *149*, 113233.

Feng, X., Lau, F. C., & Gao, D. (2009). A new bio-inspired approach to the traveling salesman problem. In *Complex Sciences* (pp. 1310–1321). Springer Berlin Heidelberg.

Ferdousy, E., Islam, M. & Matin, A. (2013). Combination of Na¨ıve Bayes Classifier and K-Nearest Neighbor (cNK) in the Classification Based Predictive Models. *Computer and Information Science*, *6*(3).

Firte, L., Lemnaru, C., & Potolea, R. (2010, August). Spam detection filter using KNN algorithm and resampling. In *International Conference on Intelligent Computer Communication and Processing* (pp. 27-33). IEEE. 10.1109/ICCP.2010.5606466

Floridi, L. (Ed.). (2010). *The Cambridge handbook of information and computer ethics*. Cambridge University Press. doi:10.1017/CBO9780511845239

Freund, Y., & Schapire, R. E. (1996). Experiments with a new boosting algorithm. In *Proceedings of the 13th International Conference on Machine Learning,* (pp. 148-156).

Fuller, T. K., Mech, L. D., & Cockrane, J. F. (2003). Wolf population dynamics. In L. D. Mech & L. Boitani (Eds.), *Wolves: Behavior, Ecology and Conservation* (pp. 161–191). University of Chicago Press.

Fumera, G., Pillai, I., Roli, F., & Biggio, B. (2007, March). Image spam filtering using textual and visual information. In *Proceedings of the MIT Spam Conference.*

Gandomi, A. H., Yang, X. S., & Alavi, A. H. (2013). Cuckoo search algorithm: A metaheuristic approach to solve structural optimization problems. *Engineering with Computers*, *29*(1), 17–35.

Ganguly, S. (2020). Multi-objective distributed generation penetration planning with load model using particle swarm optimization. *Decis. Mak. Appl. Manag. Eng.*, *3*(1), 30–42. doi:10.31181/dmame2003065g

Garcıa, J., & Fernández, F. (2015). A comprehensive survey on safe reinforcement learning. *Journal of Machine Learning Research*, *16*(1), 1437–1480.

Garg, P., & Girdhar, N. (2021, January). A Systematic Review on Spam Filtering Techniques based on Natural Language Processing Framework. In *11th International Conference on Cloud Computing, Data Science & Engineering (Confluence)* (pp. 30-35). IEEE. 10.1109/Confluence51648.2021.9377042

Gaurav, D., Tiwari, S. M., Goyal, A., Gandhi, N., & Abraham, A. (2020). Machine intelligence-based algorithms for spam filtering on document labeling. *Soft Computing*, *24*(13), 9625–9638. doi:10.100700500-019-04473-7

Ghotbi, N., Ho, M. T., & Mantello, P. (2022). Attitude of college students towards ethical issues of artificial intelligence in an international university in Japan. *AI & Society*, *37*(1), 283–290. doi:10.100700146-021-01168-2

Glasgow, R. E., Vinson, C., Chambers, D., Khoury, M. J., Kaplan, R. M., & Hunter, C. (2012). National Institutes of Health approaches to dissemination and implementation science: Current and future directions. *American Journal of Public Health*, *102*(7), 1274–1281. doi:10.2105/AJPH.2012.300755 PMID:22594758

Goh, G. L., Dikshit, V., Koneru, R., Peh, Z. K., Lu, W., Goh, G. D., & Yeong, W. Y. (2022). Fabrication of design-optimized multifunctional safety cage with conformal circuits for drone using hybrid 3D printing technology. *International Journal of Advanced Manufacturing Technology, 120*(3-4), 2573–2586. doi:10.100700170-022-08831-y

Goldberg, D. E. (1989). *Genetic Algorithms in Search, Optimization, and Machine Learning.* Addison-Wesley.

Goodfellow, I., Bengio, Y., Courville, A., & Bengio, Y. (2016). Deep learning: Vol. 1. *No. 2.* MIT Press.

Gorla, N., Somers, T. M., & Wong, B. (2010). Organizational impact of system quality, information quality, and service quality. *The Journal of Strategic Information Systems, 19*(3), 207–228. doi:10.1016/j.jsis.2010.05.001

Grassé, P. P. (1967). Nouvelles experiences sur le termite de Müller (Macrotermes mülleri) et considerations sur la théorie de la stigmergie. *Insectes Sociaux, 14*(1), 73–101.

Grier, C., Thomas, K., Paxson, V., & Zhang, M. (2010, October). @ spam: the underground on 140 characters or less. In *Proceedings of the 17th ACM conference on Computer and communications security* (pp. 27-37). ACM 10.1145/1866307.1866311

Gudigar, A., Raghavendra, U., Nayak, S., Ooi, C. P., Chan, W. Y., Gangavarapu, M. R., Dharmik, C., Samanth, J., Kadri, N. A., Hasikin, K., Barua, P. D., Chakraborty, S., Ciaccio, E. J., & Acharya, U. R. (2021). Role of Artificial Intelligence in COVID-19 Detection. *Sensors, 21*(23), 8045. https://doi.org/10.3390/s21238045.

Guo, R., Zhao, Y., Zou, Q., Fang, X., & Peng, S. (2018). Bioinformatics applications on apache spark. *GigaScience, 7*(8), giy098. PMID:30101283

Gupta, A., Verma, A., & Pramanik, S. (2022). Security Aspects in Advanced Image Processing Techniques for COVID-19. In S. Pramanik, A. Sharma, S. Bhatia, & D. N. Le (Eds.), *An Interdisciplinary Approach to Modern Network Security.* CRC Press.

Hacibeyoglu, M., Arslan, A. & Kahraman, S. (2011). Improving Classification Accuracy with Discretization on Datasets Including Continuous Valued Features. *International Scholarly and Scientific Research & Innovation, 5*(6).

Haleem, A., Javaid, M., & Vaishya, R. (2020). Effects of COVID-19 pandemic in daily life. *Current Medicine Research and Practice, 10*(2), 78.

Harmer, J. (2001). *The practice of English language teaching* .Pearson.

Harun, A., & Alam, N. (2015). Predicting Outcome of Thoracic Surgery by Data Mining Techniques. *International Journal of Advanced Research in Computer Science and Software Engineering, 5*(1).

Hasan, S., Shamsuddin, S., & Yusob, B. (2010). Enhanced Self Organizing Map (SOM) and Particle Swarm Optimization (PSO) for Classification. *Jurnal Generic, 5,* 7–11.

Hashim, F. A., Hussain, K., Houssein, E. H., Mabrouk, M. S., & Al-Atabany, W. (2021). Archimedes optimization algorithm: A new metaheuristic algorithm for solving optimization problems. *Applied Intelligence*, *51*(3), 1531–1551. doi:10.100710489-020-01893-z

Hassan, M. U., Rehmani, M. H., & Chen, J. (2019). Differential privacy techniques for cyber physical systems: A survey. *IEEE Communications Surveys and Tutorials*, *22*(1), 746–789. doi:10.1109/COMST.2019.2944748

Haupt, R. L., & Haupt, S. E. (1998). *Practical Genetic Algorithms*. Wiley Interscience.

Hew, K. F., & Brush, T. (2007). Integrating technology into K-12 teaching and learning: Current knowledge gaps and recommendations for future research. *Educational Technology Research and Development*, *55*(3), 223–252. doi:10.100711423-006-9022-5

Hicklin, J. F. (1986). Application of the genetic algorithm to automatic program generation.

Hofstadter, D. (2013). *Alan Turing: Life and legacy of a great thinker* (C. Teuscher, Ed.). Springer Science & Business Media.

Holland, J. H. (1973). Genetic algorithms and the optimal allocation of trials. *SIAM Journal on Computing*, *2*(2), 88–105. doi:10.1137/0202009

Holland, J. H. (1975). *Adaptation in natural and artificial systems*. The University of Michigan Press.

Holzinger, A., Biemann, C., Pattichis, C. S., & Kell, D. B. (2017). What do we need to build explainable AI systems for the medical domain?

Hong, J., Sapp, B., & Philbin, J. (2019). Rules of the road: Predicting driving behavior with a convolutional model of semantic interactions. In *Proceedings of the IEEE/CVF Conference on Computer Vision and Pattern Recognition* (pp. 8454-8462). 10.1109/CVPR.2019.00865

Hotchkiss, K. M., & Sampson, J. H. (2021). Temozolomide treatment outcomes and immunotherapy efficacy in brain tumor. *Journal of Neuro-Oncology*, *151*(1), 55–62. doi:10.100711060-020-03598-2 PMID:32813186

Hu, W., Gao, J., Wang, Y., Wu, O., & Maybank S.J. (2014). Online Adaboost Based Parameterized Methods for Dynamic Distributed Network Intrusion Detection. *IEEE T. Cybernetics*, (pp. 66-82).

Hu, H., Wang, H., Bai, Y., & Liu, M. (2019). Determination of endometrial carcinoma with gene expression based on optimized Elman neural network. *Applied Mathematics and Computation*, *341*, 204–214. doi:10.1016/j.amc.2018.09.005

Hu, K. (2021). Become Competent in Generating RNA-Seq Heat Maps in One Day for Novices Without Prior R Experience. In K. Hu (Ed.), *Nuclear Reprogramming. Methods in Molecular Biology* (Vol. 2239). Humana. doi:10.1007/978-1-0716-1084-8_17

Hutchinson, T., & Waters, A. (1987). *English for specific purposes*. Cambridge university press. doi:10.1017/CBO9780511733031

Hyland, K., & Hamp-Lyons, L. (2002). EAP: Issues and directions. *Journal of English for Academic Purposes*, *1*(1), 1–12. doi:10.1016/S1475-1585(02)00002-4

Imambi, S., Prakash, K. B., & Kanagachidambaresan, G. R. (2021). PyTorch. In K. B. Prakash & G. R. Kanagachidambaresan (Eds.), *Programming with TensorFlow. EAI/Springer Innovations in Communication and Computing*. Springer. doi:10.1007/978-3-030-57077-4_10

Jakhar, D., & Kaur, I. (2020). Artificial intelligence, machine learning and deep learning: Definitions and differences. *Clinical and Experimental Dermatology*, *45*(1), 131–132. doi:10.1111/ced.14029 PMID:31233628

Janecek, A., & Tan, Y. (2011, July). Iterative improvement of the multiplicative update nmf algorithm using nature-inspired optimization. In Natural Computation (ICNC), 2011 Seventh International Conference on (Vol. 3, pp. 1668-1672). IEEE. doi:10.1109/ICNC.2011.6022356

Jaya, R. S(2019) 5 chatbot ideas business cant afford to miss in 2019. *Hackernoon*. https://hackernoon.com/5-chatbots-ideas-business-cant-afford-to-miss-in-2019-e59cd41e01f8?source=rss----3a8144eabfe3---4

Jayasingh, R., Kumar, J., Telagathoti, B., Sagayam, K. M., & Pramanik, S. (2022). Speckle noise removal by SORAMA segmentation in Digital Image Processing to facilitate precise robotic surgery. *International Journal of Reliable and Quality E-Healthcare*, *11*(1), 2022. doi:10.4018/IJRQEH.295083

Jeyaraj, A. (2020). DeLone & McLean models of information system success: Critical meta-review and research directions. *International Journal of Information Management*, *54*, 102139. doi:10.1016/j.ijinfomgt.2020.102139

Jing, J., Liu, S., Wang, G., Zhang, W., & Sun, C. (2022). Recent advances on image edge detection: A comprehensive review. *Neurocomputing*, *503*, 259–271. doi:10.1016/j.neucom.2022.06.083

Joe, T. (2021). Deep Learning and ML in Health Care: Use Cases, Examples. *TopFlight*. https://topflightapps.com/ideas/machine-learning-in-healthcare-use-cases/

K.aushik, D., Garg, V., Annu, Gupta, A. & Pramanik, S. (2021). Application of Machine Learning and Deep Learning in Cyber security: An Innovative Approach. In M. Ghonge, S. Pramanik, R. Mangrulkar and D. N. Le, (Eds.), *Cybersecurity and Digital Forensics: Challenges and Future Trends*. Wiley.

Kaelber, D. C., Jha, A. K., Johnston, D., Middleton, B., & Bates, D. W. (2008). A research agenda for personal health records (PHRs). *Journal of the American Medical Informatics Association*, *15*(6), 729–736. doi:10.1197/jamia.M2547 PMID:18756002

Kalchbrenner, N., Grefenstette, E., & Blunsom, P. (2014). *A convolutional neural network for modelling sentences.* arXiv preprint arXiv:1404.2188.

Kanan, T., Mughaid, A., Al-Shalabi, R., Al-Ayyoub, M., Elbes, M., & Sadaqa, O. (2022). Business intelligence using deep learning techniques for social media contents. *Cluster Computing*. Advance online publication. doi:10.100710586-022-03626-y

Karunarathne, S. M., Saxena, N., & Khan, M. K. (2021). Security and privacy in IoT smart healthcare. *IEEE Internet Computing*, *25*(4), 37–48. doi:10.1109/MIC.2021.3051675

Katoch, S., Chauhan, S. S., & Kumar, V. (2021). A review on genetic algorithm: Past, present, and future. *Multimedia Tools and Applications*, *80*(5), 8091–8126. doi:10.100711042-020-10139-6 PMID:33162782

Kaur, P. D., & Chana, I. (2014). Cloud-based intelligent system for delivering health care as a service. *Computer Methods and Programs in Biomedicine*, *113*(1), 346–359. doi:10.1016/j.cmpb.2013.09.013 PMID:24139021

Kaushik, D., & Garg, M. Annu, Gupta, A. and Pramanik, S. (2021). Application of Machine Learning and Deep Learning in Cyber security: An Innovative Approach. In M. Ghonge, S. Pramanik, R. Mangrulkar and D. N. Le, (eds.), Cybersecurity and Digital Forensics: Challenges and Future Trends. Wiley.

Kayed, M., Anter, A., & Mohamed, H. (2020, February). Classification of garments from fashion mnist dataset using cnn lenet-5 architecture. In *2020 International Conference on Innovative Trends in Communication and Computer Engineering (ITCE)* (pp. 238-243). IEEE. 10.1109/ITCE48509.2020.9047776

Kennedy, J., & Eberhart, R. ((1995). Particle swarm optimization. *In Proceedings International Conference on Neural Networks,* (pp. 1942-1948).

Kennedy, J., & Eberhart, R. (1995). Particle Swarm Optimization. *Proceedings of IEEE International Conference on Neural Networks*, pp. 1942–1948 10.1109/ICNN.1995.488968

Kennedy, J., & Eberhart, R. C. Particle Swarm Optimization. In *Proceedings of IEEE International Conference on Neural Networks*, Perth, Australia, pp. 1942–1948, 1995

Keshta, I., & Odeh, A. (2021). Security and privacy of electronic health records: Concerns and challenges. *Egyptian Informatics Journal*, *22*(2), 177–183. doi:10.1016/j.eij.2020.07.003

Khan, I., Dewangan, B., Meena, A., & Birthare, M. (2020, March). Study of Various Cloud Service Providers: A Comparative Analysis. At *the 5th International Conference on Next Generation Computing Technologies*, 1-5. 10.2139srn.3672950

Kinnear, K. E. (1994). A Perspective on the Work in this Book. K. E. Kinnear (Ed.), Advances in Genetic Programming (pp. 3-17). Cambridge: MIT Press.

Ko, D. H., Choi, S. H., Shin, J. M., Liu, P., & Choi, Y. H. (2020). Structural image de-identification for privacy-preserving deep learning. *IEEE Access: Practical Innovations, Open Solutions*, *8*, 119848–119862. doi:10.1109/ACCESS.2020.3005911

Koren, Y., Rendle, S., & Bell, R. (2022). Advances in Collaborative Filtering. In F. Ricci, L. Rokach, & B. Shapira (Eds.), *Recommender Systems Handbook.* Springer. doi:10.1007/978-1-0716-2197-4_3

Koza, J. R. (1994). Genetic programming II: Automatic discovery of reusable subprograms. Cambridge, MA, USA.

Koza, J. R. (1992). *Genetic programming: on the programming of computers by means of natural selection* (Vol. 1). MIT press.

Koza, J. R. (1994). Introduction to Genetic Programming. In K. E. Kinnear (Ed.), *Advances in Genetic Programming* (pp. 21–41). MIT Press.

Krasser, S., Tang, Y., Gould, J., Alperovitch, D., & Judge, P. (2007, June). Identifying image spam based on header and file properties using C4. 5 decision trees and support vector machine learning. In *Information Assurance and Security Workshop* (pp. 255-261). IEEE.

Krenker, A., Bešter, J., & Kos, A. (2011). Introduction to the artificial neural networks. In *Artificial Neural Networks: Methodological Advances and Biomedical Applications.* InTech.

Krittanawong, C., Johnson, K. W., Choi, E., Kaplin, S., Venner, E., Murugan, M., & Tang, W. W. (2022). Artificial Intelligence and Cardiovascular Genetics. *Life (Chicago, Ill.)*, *12*(2), 279. PMID:35207566

Krittanawong, C., Zhang, H., Wang, Z., Aydar, M., & Kitai, T. (2017). Artificial intelligence in precision cardiovascular medicine. *Journal of the American College of Cardiology*, *69*(21), 2657–2664. doi:10.1016/j.jacc.2017.03.571 PMID:28545640

Kumar, A., & Kumar, R. (2019, March). Privacy-Preservation of Vertically Partitioned Electronic Health Record using Perturbation Methods. In *6th International Conference on Computing for Sustainable Global Development (INDIACom)* (pp. 161-166). IEEE.

Lapointe, S., Perry, A., & Butowski, N. A. (2018). Primary brain tumours in adults. *Lancet*, *392*(10145), 432–446. doi:10.1016/S0140-6736(18)30990-5 PMID:30060998

Lee, J. H., Kim, D. H., Jeong, S. N., & Choi, S. H. (2018). Detection and diagnosis of dental caries using a deep learning-based convolutional neural network algorithm. *Journal of Dentistry*, *77*, 106–111. doi:10.1016/j.jdent.2018.07.015 PMID:30056118

Leevy, J. L., Khoshgoftaar, T. M., & Villanustre, F. (2020). Survey on RNN and CRF models for de-identification of medical free text. *Journal of Big Data*, *7*(1), 1–22. doi:10.118640537-020-00351-4

Leisink, P., & Boxall, P. (2021). HRM and social legitimacy in the public sector. In *Research Handbook on HRM in the Public Sector*. Edward Elgar Publishing.

Li, J., Zheng, S., & Tan, Y. (2014, July). Adaptive Fireworks Algorithm. InEvolutionary Computation (CEC), 2014 IEEE Congress on (pp. 3214-3221). IEEE. doi:10.1109/CEC.2014.6900418

Li, M., & Chapman, G. B. (2020). Medical decision-making. The Wiley Encyclopedia of Health Psychology, 347-353.

Lin, J., Li, Y., & Yang, G. (2021). FPGAN: Face de-identification method with generative adversarial networks for social robots. *Neural Networks*, *133*, 132–147. doi:10.1016/j.neunet.2020.09.001 PMID:33217682

Liu, C., Yan, X., Liu, C., & Wu, H. (2011). The wolf colony algorithm and its application. *Chinese Journal of Electronics*, *20*, 212–216.

Liu, J., Zheng, S., & Tan, Y. (2013). The improvement on controlling exploration and exploitation of firework algorithm. In *Advances in swarm intelligence* (pp. 11–23). Springer Berlin Heidelberg., doi:10.1007/978- 3-642-38703-6_2

Liu, Z., Mao, H., Wu, C. Y., Feichtenhofer, C., Darrell, T., & Xie, S. (2022). A convnet for the 2020s. In *Proceedings of the IEEE/CVF Conference on Computer Vision and Pattern Recognition* (pp. 11976-11986).

Li, W.-W., Wang, H., & Zou, Z. J. (2005). Function optimization method based on bacterial colony chemotaxis. *Journal of Circuits and Systems*, *10*, 58–63.

Li, Y., & Du, Y. (2019). A novel software-defined convolutional neural networks accelerator. *IEEE Access: Practical Innovations, Open Solutions*, *7*, 177922–177931. doi:10.1109/ACCESS.2019.2956841

Longoni, C., Bonezzi, A., & Morewedge, C. K. (2019). Resistance to artificial medical intelligence. *The Journal of Consumer Research*, *46*(4), 629–650. doi:10.1093/jcr/ucz013

Lou, S., Pagani, L., Zeng, W., Jiang, X., & Scott, P. J. (2020). Watershed segmentation of topographical features on freeform surfaces and its application to additively manufactured surfaces. *Precision Engineering*, *63*, 177–186. doi:10.1016/j.precisioneng.2020.02.005

Luengo-Oroz, M., Hoffmann Pham, K., Bullock, J., Kirkpatrick, R., Luccioni, A., Rubel, S., Wachholz, C., Chakchouk, M., Biggs, P., Nguyen, T., Purnat, T., & Mariano, B. (2020). Artificial intelligence cooperation to support the global response to COVID-19. *Nature Machine Intelligence*, *2*(6), 295–297. doi:10.103842256-020-0184-3

Macdonald, D. W., Creel, S., & Mills, M. G. L. (2004). Society: Canid society. In D. W. Macdonald & C. Sillero-Zubiri (Eds.), *Biology and Conservation of Wild Carnivores* (pp. 85–106). Oxford University Press.

Madanian, S., Nguyen, H. H., & Mirza, F. (2021). Wearable Technology. In D. Gu & M. E. Dupre (Eds.), *Encyclopedia of Gerontology and Population Aging*. Springer. doi:10.1007/978-3-030-22009-9_459

Maedche, A., Legner, C., Benlian, A., Berger, B., Gimpel, H., Hess, T., Hinz, O., Morana, S., & Söllner, M. (2019). AI-based digital assistants. *Business & Information Systems Engineering*, *61*(4), 535–544. doi:10.100712599-019-00600-8

Makridis, C., & Mishra, S. (2020) How Does Artificial Intelligence Raise Well-being? Evidence from Cities in the United States.: https://ssrn.com/abstract=3669348 doi:10.2139/ssrn.3669348

Malik, H., Fatema, N., & Alzubi, J. A. (Eds.). (2021). *AI and machine learning paradigms for health monitoring system: intelligent data analytics*. Springer. doi:10.1007/978-981-33-4412-9

Maly, V., Maly, O., Kolostova, K., & Bobek, V. (2019). Circulating tumor cells in diagnosis and treatment of lung cancer. *In Vivo, 33*(4), 1027-1037.

Mandal, A., Dutta, S., & Pramanik, S. (2021). Machine Intelligence of Pi from Geometrical Figures with Variable Parameters using SCILab. In D. Samanta, R. R. Althar, S. Pramanik, & S. Dutta (Eds.), *Methodologies and Applications of Computational Statistics for Machine Learning* (pp. 38–63). IGI Global., doi:10.4018/978-1-7998-7701-1.ch003

Mani, S. (2019). *Artistic Enhancement and Style Transfer of Image Edges using Directional Pseudo-coloring*. arXiv preprint arXiv:1906.07981.

Mansour, R. F., & Aljehane, N. O. (2021). An optimal segmentation with deep learning based inception network model for intracranial hemorrhage diagnosis. *Neural Computing & Applications, 33*(20), 13831–13843. doi:10.100700521-021-06020-8

Mason, R. O. (2017). Four ethical issues of the information age. In *Computer ethics* (pp. 41–48). Routledge. doi:10.4324/9781315259697-8

Mateen, M., Wen, J., Song, S., & Huang, Z. (2018). Fundus image classification using VGG-19 architecture with PCA and SVD. *Symmetry, 11*(1), 1. doi:10.3390ym11010001

Matrix. (n.d.). Neuro-inspired, AI: The UTSA AI Consortium for Human Well-Being. *Matrix*. https://ai.utsa.edu/research/neuro-inspired-artificial-intelligence/

Mayes, J. L. (2020). *UDL and Motivation: Student Perceptions of the Impact of Universal Design for Learning on Motivation of First-Year Community College Students in Rural East Tennessee* [Doctoral dissertation, East Tennessee State University, United States].

Mayorga-Ruiz, I., Jiménez-Pastor, A., Fos-Guarinos, B., López-González, R., García-Castro, F., & Alberich-Bayarri, Á. (2019). The Role of AI in Clinical Trials. In *Artificial Intelligence in Medical Imaging* (pp. 231–243). Springer. doi:10.1007/978-3-319-94878-2_16

Meslie, Y., Enbeyle, W., Pandey, B. K., Pramanik, S., Pandey, D., Dadeech, P., Belay, A., & Saini, A. (2021). Machine Intelligence-based Trend Analysis of COVID-19 for Total Daily Confirmed Cases in Asia and Africa. In D. Samanta, R. R. Althar, S. Pramanik, & S. Dutta (Eds.), *Methodologies and Applications of Computational Statistics for Machine Learning* (pp. 164–185). IGI Global., doi:10.4018/978-1-7998-7701-1.ch009

Michael, A. (2009). *Armstrong's handbook of human resource management practice*.

Minaee, S., Boykov, Y. Y., Porikli, F., Plaza, A. J., Kehtarnavaz, N., & Terzopoulos, D. (2021). Image segmentation using deep learning: A survey. *IEEE Transactions on Pattern Analysis and Machine Intelligence*, 1. doi:10.1109/TPAMI.2021.3059968 PMID:33596172

Mirjalili, S. (2015). The Ant Lion Optimizer. *Advances in Engineering Software*, *83*, 80–98. doi:10.1016/j.advengsoft.2015.01.010

Mirjalili, S., Mirjalili, S. M., & Lewis, A. (2014). Grey Wolf Optimizer. *Advances in Engineering Software*, *69*, 46–61. doi:10.1016/j.advengsoft.2013.12.007

Mitchell, M. (1995). Genetic Algorithms: An Overview. *Complexity*, *1*(1), 31–39. doi:10.1002/cplx.6130010108

Mitchell, M. (1998). *An introduction to genetic algorithms*. MIT press.

Mohammad, A. H., & Zitar, R. A. (2011). Application of genetic optimized artificial immune system and neural networks in spam detection. *Applied soft computing, 11*(4), 3827-3845.

Monova-Zheleva, M., Zhelev, Y., & Nikolova, E. (2021). Fostering technology-enhanced learning and digital innovations in schools–key factors and challenges. *Mathematics and Education in Mathematics*, *50*, 215–221.

Monroe, C., Tazi, F., & Das, S. (2021). Location data and COVID-19 contact tracing: how data privacy regulations and cell service providers work in tandem. *Researchgate*.

Morley, J., Floridi, L., Kinsey, L., & Elhalal, A. (2020). From what to how: An initial review of publicly available AI ethics tools, methods, and research to translate principles into practices. *Science and Engineering Ethics*, *26*(4), 2141–2168. doi:10.100711948-019-00165-5 PMID:31828533

Musen, M. A., Middleton, B., & Greenes, R. A. (2021). Clinical decision-support systems. In *Biomedical informatics* (pp. 795–840). Springer. doi:10.1007/978-3-030-58721-5_24

Natalie, T. (2021) ML in Health-care: 12 Real-World Use Cases to Know. *NixUnited* https://nix-united.com/blog/machine-learning-in-healthcare-12-real-world-use-cases-to-know/

Negi, G., Kumar, A., Pant, S., & Ram, M. (2021). Optimization of Complex System Reliability using Hybrid Grey Wolf Optimizer. *Decis. Mak. Appl. Manag. Eng.*, *4*(2), 241–256. doi:10.31181/dmame210402241n

Nissenbaum, H. (2020). Protecting privacy in an information age: The problem of privacy in public. In *The Ethics of Information Technologies* (pp. 141–178). Routledge. doi:10.4324/9781003075011-12

Niu, B., & Wang, H. (2012). *Bacterial colony optimization*. Discrete Dynamics in Nature and.

Niu, X., Song, T., & Xiong, H. (2021). Large scale synthesis of red emissive carbon dots powder by solid state reaction for fingerprint identification. *Chinese Chemical Letters*, *32*(6), 1953–1956. doi:10.1016/j.cclet.2021.01.006

Nwankpa, C., Ijomah, W., Gachagan, A., & Marshall, S. (2018). Activation functions: Comparison of trends in practice and research for deep learning. *Cornell University ARIV.*

Nyangaresi, V. O. (2022). A Formally Validated Authentication Algorithm for Secure Message Forwarding in Smart Home Networks. *SN. Computer Science*, *3*, 364. doi:10.100742979-022-01269-9

O'Shea, R. J., Sharkey, A. R., Cook, G. J., & Goh, V. (2021). Systematic review of research design and reporting of imaging studies applying convolutional neural networks for radiological cancer diagnosis. *European Radiology*, *31*(10), 7969–7983. doi:10.100700330-021-07881-2 PMID:33860829

on bacterial evolution and its applications for scheduling a flexible manufacturing system. Jurnal Teknik Industri, 14, 1–12

Osoba, O. A., & Welser, I. V. W. (2017). *An intelligence in our image: The risks of bias and errors in artificial intelligence*. Rand Corporation.

Palmer, J. D. (1978). English for Specific Purposes: A Case Study Approach.

Passino, K. M. (2012). Bacterial foraging optimization. Innovations and Developments of Swarm Intelligence Applications, 219.

Pei, Y., Zheng, S., Tan, Y., & Takagi, H. (2012, October). An empirical study on influence of approximation approaches on enhancing fireworks algorithm. In Systems, Man, and Cybernetics (SMC), 2012 IEEE International Conference on (pp. 1322-1327). IEEE. doi:10.1109/ICSMC.2012.6377916

Pohlheim, H. (2005). Evolutionary algorithms: overview, methods and operators. GEATbx: Gentic & Evolutionary AlgorithmToolbox for Matlab.

Poli, R., Langdon, W. B., McPhee, N. F., & Koza, J. R. (2008). A field guide to genetic programming. Lulu. com.

Prabu, A. (2021). SmartScope: An AI-Powered Digital Auscultation Device To Detect Cardiopulmonary Diseases.

Pramanik, S., Joardar, S., Jena, O. P., & Obaid, A. J. (2022). An Analysis of the Operations and Confrontations of Using Green IT in Sustainable Farming. *2nd International Conference on Modern Applications of Information and Communication Technology*, Al-Kadhum, Iraq.

Pramanik, S., Sagayam, K. M., & Jena, O. P. (2021). Machine Learning Frameworks in Cancer Detection. ICCSRE 2021, Morocco, North Africa.

Pramanik, S. (2022). An Effective Secured Privacy-Protecting Data Aggregation Method in IoT. In M. O. Odhiambo, W. Mwashita, & I. G. I. Global (Eds.), *Achieving Full Realization and Mitigating the Challenges of the Internet of Things.*, doi:10.4018/978-1-7998-9312-7.ch008

Pramanik, S. (2022). Carpooling Solutions using Machine Learning Tools. In *Handbook of Research on Evolving Designs and Innovation in ICT and Intelligent Systems for Real-World Applications, K. K. Sarma, N. Saikia and M. Sharma*. IGI Global. doi:10.4018/978-1-7998-9795-8.ch002

Pramanik, S., & Bandyopadhyay, S. (2022). Analysis of Big Data. In I. G. I. John Wang (Ed.), *Encyclopedia of Data Science and Machine Learning*. Global.

Pramanik, S., & Bandyopadhyay, S. (2022). Identifying Disease and Diagnosis in Females using Machine Learning. In I. G. I. John Wang (Ed.), *Encyclopedia of Data Science and Machine Learning*. Global.

Pramanik, S., Galety, M. G., Samanta, D., & Joseph, N. P. (2022). Data Mining Approaches for Decision Support Systems. *3rd International Conference on Emerging Technologies in Data Mining and Information Security.*

Pramanik, S., Singh, R. P., & Ghosh, R. (2020). Application of Bi-orthogonal Wavelet Transformand Genetic Algorithm in Image Steganography. *Multimedia Tools and Applications*, *79*(25-26), 17463–17482. doi:10.100711042-020-08676-1

Precup, R.-E., David, R.-C., Roman, R.-C., Petriu, E. M., & Szedlak-Stinean, A.-I. (2021). Slime Mould Algorithm-Based Tuning of Cost-Effective Fuzzy Controllers for Servo Systems. *Int. J. Comput. Intell. Syst.*, *14*(1), 1042–1052. doi:10.2991/ijcis.d.210309.001

Premaratne, U., Samarabandu, J., & Sidhu, T. (2009, December). A new biologically inspired optimization algorithm. In Industrial and Information Systems (ICIIS), 2009 International Conference on (pp. 279-284). IEEE.

Priyadharsini, R., & Sharmila, T. S. (2019). Object detection in underwater acoustic images using edge based segmentation method. *Procedia Computer Science*, *165*, 759–765. doi:10.1016/j.procs.2020.01.015

Qian, X., Tang, Y. Y., Yan, Z., & Hang, K. (2013). ISABoost: A weak classifier inner structure adjusting based AdaBoost algorithm-ISABoost based application in scene categorization. *Neurocomputing 103,* (pp. 104-113). Elsevier.

Qiu, X., & Lau, H. (2014). An AIS-based hybrid algorithm for static job shop scheduling problema. *Journal of Intelligent Manufacturing*, *25*(3), 489–503. doi:10.100710845-012-0701-2

Rai, V., Quang, D. X., Erdos, M. R., Cusanovich, D. A., Daza, R. M., Narisu, N., Zou, L. S., Didion, J. P., Guan, Y., Shendure, J., Parker, S. C. J., & Collins, F. S. (2020). Single-cell ATAC-Seq in human pancreatic islets and deep learning upscaling of rare cells reveals cell-specific type 2 diabetes regulatory signatures. *Molecular Metabolism*, *32*, 109–121. doi:10.1016/j.molmet.2019.12.006 PMID:32029221

Ramageri, B. M. (2010). Data Mining Techniques and Applications. *Indian Journal of Computer Science and Engineering*, *1*(4), 301–305.

Ramesh, K. K. D., Kumar, G. K., Swapna, K., Datta, D., & Rajest, S. S. (2021). A review of medical image segmentation algorithms. *EAI Endorsed Transactions on Pervasive Health and Technology*, *7*(27), e6–e6.

Rapacz, S., Chołda, P., & Natkaniec, M. (2021). A Method for Fast Selection of Machine-Learning Classifiers for Spam Filtering. *Electronics (Basel)*, *10*(17), 2083. doi:10.3390/electronics10172083

Rebello, R. J., Oing, C., Knudsen, K. E., Loeb, S., Johnson, D. C., Reiter, R. E., Gillessen, S., Van der Kwast, T., & Bristow, R. G. (2021). Prostate cancer. *Nature Reviews. Disease Primers*, *7*(1), 1–27. doi:10.103841572-020-00243-0 PMID:33542230

Rechenberg, I. (1989). *Evolution strategy: Nature's way of optimization. InOptimization: Methods and applications, possibilities and limitations*. Springer Berlin Heidelberg.

Reeves, C. (2003). *Genetic algorithms.*

Reynolds, C. W. (1987, August). Flocks, herds and schools: A distributed behavioral model. []. ACM.]. *Computer Graphics*, *21*(4), 25–34.

Richters, M. M., Xia, H., Campbell, K. M., Gillanders, W. E., Griffith, O. L., & Griffith, M. (2019). Best practices for bioinformatic characterization of neoantigens for clinical utility. *Genome Medicine*, *11*(1), 1–21. doi:10.118613073-019-0666-2 PMID:31462330

Runkler, T. A. (2008). Wasp swarm optimization of the c-means clustering model. *International Journal of Intelligent Systems*, *23*(3), 269–285.

Saberi, A., Vahidi, M., & Bidgoli, B. M. (2007, November). Learn to detect phishing scams using learning and ensemble? methods. In *Web Intelligence and Intelligent Agent Technology Workshops, 2007 IEEE/WIC/ACM International Conferences*, (pp. 311-314). IEEE.

Sahiner, B., Pezeshk, A., Hadjiiski, L. M., Wang, X., Drukker, K., Cha, K. H., Summers, R. M., & Giger, M. L. (2019). Deep learning in medical imaging and radiation therapy. *Medical Physics*, *46*(1), e1–e36. doi:10.1002/mp.13264 PMID:30367497

Sakshi, G. S., Sharma, C., Kukreja, V. (2021). Handwritten Mathematical Symbols Classification Using WEKA. In Choudhary, A., Agrawal, A.P., Logeswaran, R., Unhelkar, B. (eds) Applications of Artificial Intelligence and Machine Learning. Lecture Notes in Electrical Engineering (vol. 778). Springer, Singapore. doi:10.1007/978-981-16-3067-5_4

Samanta, D., Dutta, S., Galety, M. G., & Pramanik, S. (2021). A Novel Approach for Web Mining Taxonomy for High-Performance Computing, The *4th International Conference of Computer Science and Renewable Energies (ICCSRE'2021).* doi:10.1051/e3sconf/202129701073

Sanakal, R., & Jayakumari, S. T. (2014). Prognosis of Diabetes Using Data mining Approach-Fuzzy C Means Clustering and Support Vector Machine. *International Journal of Computer Trends and Technology*, *11*(2), 94–98. doi:10.14445/22312803/IJCTT-V11P120

Schapire, R. E. (1990). The Strength of Weak Learnability. *Machine Learning*, *5*(2), 197–227. doi:10.1007/BF00116037

Schloemer, T. H. (2022). Managing big data. *Nature Energy*, *7*(2), 122–123. doi:10.103841560-021-00972-w

Schudel, A., Francis, D. M., & Thomas, S. N. (2019). Material design for lymph node drug delivery. *Nature Reviews. Materials*, *4*(6), 415–428.

Shahab, S., Agarwal, P., Mufti, T., & Obaid, A. J. (2022). SIoT (Social Internet of Things): A Review. In S. Fong, N. Dey, & A. Joshi (Eds.), *ICT Analysis and Applications. Lecture Notes in Networks and Systems* (Vol. 314). Springer. doi:10.1007/978-981-16-5655-2_28

Sharma, S. (2017). Activation functions in neural networks. *Towards Data Science, 6*.

Sharma, N., Sharma, R., & Jindal, N. (2021). Machine learning and deep learning applications-a vision. *Global Transitions Proceedings*, *2*(1), 24–28. doi:10.1016/j.gltp.2021.01.004

Shim, M., & Jo, H. S. (2020). What quality factors affect the perceived benefits of online health information sites? Application of the updated DeLone and McLean Information Systems Success Model. *International Journal of Medical Informatics*, *137*, 104093. doi:10.1016/j.ijmedinf.2020.104093 PMID:32078918

Shuaib, M., Alam, S., Alam, M. S., & Nasir, M. S. (2021). Compliance with HIPAA and GDPR in blockchain-based electronic health record. *Materials Today: Proceedings*. doi:10.1016/j.matpr.2021.03.059

Sindhu, V., Prabha, S., Veni, S., & Hemalatha, M. (2014). Thoracic Surgery Analysis Using Data Mining Techniques. *Int.J. Computer Technology and Application*, *5*(2), 578–586.

Singh, J., & Singh, J. (2020). COVID-19 and its impact on society. *Electronic Research Journal of Social Sciences and Humanities, 2*.

Sinha, M., Chacko, E., Makhija, P., & Pramanik, S. (2021). Energy Efficient Smart Cities with Green IoT. In C. Chakrabarty (Ed.), *Green Technological Innovation for Sustainable Smart Societies: Post Pandemic Era*. Springer. doi:10.1007/978-3-030-73295-0_16

Soliman, M., Hassanien, A., Ghali, N., & Onsi, H. (2012). An adaptive Water- marking Approach for Medical Imaging Using Swarm Intelligent. *International Journal of Smart Home*, *6*(1), 37–50.

Soni, M., & Singh, D. K. (2021). Privacy-preserving authentication and Key-Management protocol for health information systems. In *Data Protection and Privacy in Healthcare* (pp. 37–50). CRC Press. doi:10.1201/9781003048848-3

Sowmiya, B., & Poovammal, E. (2021). A heuristic K-anonymity based privacy preserving for student management hyperledger fabric blockchain. *Wireless Personal Communications*, 1–18.

Sparrow, R. (2007). Killer robots. *Journal of Applied Philosophy*, *24*(1), 62–77. doi:10.1111/j.1468-5930.2007.00346.x

Srivastava, S., Kumar, P., Chaudhry, V., & Singh, A. (2020). Detection of ovarian cyst in ultrasound images using fine-tuned VGG-16 deep learning network. *SN Computer Science*, *1*(2), 1–8. doi:10.100742979-020-0109-6

Stefaniak, J. E. (2019). Instructional design theory. In *Open and Distance Education Theory Revisited* (pp. 85–94). Springer.

Steven, Y. (2017). *Lin, Tait D Shanafelt, Steven M Asch.* Reimagining Clinical Documentation With Artificial Intelligence.

Storn, R., & Price, K. (1997). Differential evolution–a simple and efficient heuristic for global optimization over continuous spaces. *Journal of Global Optimization*, *11*(4), 341–359.

Strevens, P. (1977). Special-purpose language learning: A perspective. *Language Teaching*, *10*(3), 145–163.

Stützle, T., & Hoos, H. (1997, April). MAX-MIN ant system and local search for the traveling salesman problem. In Evolutionary Computation, 1997., IEEE International Conference on (pp. 309-314). IEEE.

Sultana, F., Sufian, A., & Dutta, P. (2018, November). Advancements in image classification using convolutional neural network. In *Fourth International Conference on Research in Computational Intelligence and Communication Networks (ICRCICN)* (pp. 122-129). IEEE. 10.1109/ICRCICN.2018.8718718

Sun, J., Palade, V., Cai, Y., Fang, W., & Wu, X. (2014). Biochemical systems identification by a random drift particle swarm optimization approach. *BMC Bioinformatics 15*(Suppl 6), 1-17. http://www.biomedcentral.com/1471- 2105/15/S6/S1.

Sundjaja, J. H., & Bordoni, B. (2021). Anatomy, Thorax, Lung Veins. StatPearls Publishing.

Supriya, M., & Chattu, V. K. (2021). A Review of Artificial Intelligence, Big Data, and Blockchain Technology Applications in Medicine and Global Health. *Big Data and Cognitive Computing.*, *5*(3), 41. doi:10.3390/bdcc5030041

Suykens, J. A., & Vandewalle, J. (1999). Least squares support vector machine classifiers. *Neural Processing Letters*, *9*(3), 293–300. doi:10.1023/A:1018628609742

Szepesvári, C. (2010). Algorithms for reinforcement learning. *Synthesis Lectures on Artificial Intelligence and Machine Learning*, *4*(1), 1-103.

Taweesiriwate, A., Manaskasemsak, B., & Rungsawang, A. (2012, March). Web spam detection using link-based ant colony optimization. In *International Conference on Advanced Information Networking and Applications (AINA)* (pp. 868-873). IEEE. 10.1109/AINA.2012.118

Timan, T., & Mann, Z. (2021). Data Protection in the Era of Artificial Intelligence: Trends, Existing Solutions and Recommendations for Privacy-Preserving Technologies. In *The Elements of Big Data Value* (pp. 153–175). Springer. doi:10.1007/978-3-030-68176-0_7

Tinio, V. L. (2002). *ICT in Education. United Nations Development Programme Bureau for development policy. Lind, A. (2008). Literacy for all: Making a difference*. Unesco.

Toreini, E., Aitken, M., Coopamootoo, K., Elliott, K., Zelaya, C. G., & Van Moorsel, A. (2020, January). The relationship between trust in AI and trustworthy machine learning technologies. In *Proceedings of the 2020 conference on fairness, accountability, and transparency* (pp. 272-283). 10.1145/3351095.3372834

Tosepu, R., Gunawan, J., Effendy, D. S., Lestari, H., Bahar, H., & Asfian, P. (2020). Correlation between weather and Covid-19 pandemic in Jakarta, Indonesia. *The Science of the Total Environment*, *725*, 138436. doi:10.1016/j.scitotenv.2020.138436 PMID:32298883

Troncoso, C., Payer, M., Hubaux, J. P., Salathé, M., Larus, J., Bugnion, E., & Pereira, J. (2020). Decentralized privacy-preserving proximity tracing.

Valdes-Ramirez, D., Medina-Pérez, M. A., Monroy, R., Loyola-González, O., Rodríguez-Ruiz, J., Morales, A., & Herrera, F. (2019). A Review of Fingerprint Feature Representations and Their Applications for Latent Fingerprint Identification: Trends and Evaluation. *IEEE Access: Practical Innovations, Open Solutions*, *7*(1), 48484–48499. doi:10.1109/ACCESS.2019.2909497

Valdez, F., Castillo, O., Cortes-Antonio, P., & Melin, P. (2020). A survey of Type-2 fuzzy logic controller design using nature inspired optimization. *Journal of Intelligent & Fuzzy Systems*, *39*(5), 6169–6179. doi:10.3233/JIFS-189087

Veatch, R. M. (2012). *Hippocratic, religious, and secular medical ethics: the points of conflict*. Georgetown University Press.

Ventola, C. L. (2014). Social media and health care professionals: Benefits, risks, and best practices. *P&T*, *39*(7), 491. PMID:25083128

Vora, S., & Mehta, R. (2012). MCAIM: Modified CAIM Discretization Algorithm for Classification. *International Journal of Applied Information Systems,* 3(5).

Wallace, A. R. (1870). *Contributions to the theory of natural selection: A series of essays*. Macmillan and Company.

Wang, Y., Li, Y., Xiong, M., & Jin, L. (2015). Random Bits Regression: a Strong General Predictor for Big Data.

Wang, S., Mathew, A., Chen, Y., Xi, L., Ma, L., & Lee, J. (2009). Empirical analysis of support vector machine ensemble classifiers. *Expert Systems with Applications*, *36*(3), 6466–6476. doi:10.1016/j.eswa.2008.07.041

Wang, T., Du, Z., Zhu, F., Cao, Z., An, Y., Gao, Y., & Jiang, B. (2020). Comorbidities and multi-organ injuries in the treatment of COVID-19. *Lancet*, *395*(10228), e52. doi:10.1016/S0140-6736(20)30558-4 PMID:32171074

Welc, J. (2022). Financial Statement Analysis. In *Evaluating Corporate Financial Performance*. Palgrave Macmillan. doi:10.1007/978-3-030-97582-1_3

Whitman, W. B., Chuvochina, M., Hedlund, B. P., Hugenholtz, P., Konstantinidis, K. T., Murray, A., Palmer, M., Parks, D. H., Probst, A. J., Reysenbach, A.-L., Rodriguez-R, L. M., Rossello-Mora, R., Sutcliffe, I., & Venter, S. N. (2022). Development of the SeqCode: A proposed nomenclatural code for uncultivated prokaryotes with DNA sequences as type. *Systematic and Applied Microbiology*, *45*(5), 126305. doi:10.1016/j.syapm.2022.126305 PMID:36049255

Wilder-Smith, A., & Freedman, D. O. (2020). Isolation, quarantine, social distancing and community containment: Pivotal role for old-style public health measures in the novel coronavirus (2019-nCoV) outbreak. *Journal of Travel Medicine*, *27*(2), 2. doi:10.1093/jtm/taaa020 PMID:32052841

Witten, I. H., & Frank, E. (2005). Data Mining: Practical Machine Learning Tools and Techniques (2nd ed.). Morgan Kaufmann Publishers, Elsevier Inc.

World Health Organization. (2020). Advice on the use of masks in the context of COVID-19: interim guidance, 5 June 2020 (No. WHO/2019-nCoV/IPC_Masks/2020.4). World Health Organization.

Wu, D., Liao, Y., Hu, C., Yu, S., & Tian, Q. (2021). An Enhanced Fuzzy Control Strategy for Low-Level Thrusters in Marine Dynamic Positioning Systems Based on Chaotic Random Distribution Harmony Search. *International Journal of Fuzzy Systems*, *23*(6), 1823–1839. doi:10.100740815-020-00989-5

Wu, J. (2011). Integrated Real-Coded Genetic Algorithm and Particle Swarm Optimization for Solving Constrained Global Optimization Problems. *Advances in Information Technology and Education Communications in Computer and Information Science*, *201*, 511–522. doi:10.1007/978-3-642-22418-8_72

Wu, J. H., & Wang, Y. M. (2006). Measuring KMS success: A respecification of the DeLone and McLean's model. *Information & Management*, *43*(6), 728–739. doi:10.1016/j.im.2006.05.002

Yaacoub, J. P. A., Noura, M., Noura, H. N., Salman, O., Yaacoub, E., Couturier, R., & Chehab, A. (2020). Securing the Internet of medical systems: Limitations, issues, and recommendations. *Future Generation Computer Systems*, *105*, 581–606. doi:10.1016/j.future.2019.12.028

Yang, M., Lyu, L., Zhao, J., Zhu, T., & Lam, K. Y. (2020). Local differential privacy and its applications: A comprehensive survey. *arXiv preprint arXiv:2008.03686.*

Yang, C., Tu, X., & Chen, J. (2007). Algorithm of marriage in honey bees optimization based on the wolf pack search. In *IEEE International Conference on Intelligent Pervasive Computing (IPC)* (pp. 462–467)

Yıkılmaz, İ. (2020). New era: The transformation from the information society to super smart society (society 5.0). In G. Mert, E. Şen, & O. Yılmaz (Eds.), *Data, Information and Knowledge Management* (pp. 85–112).

Yoshida, H., Kawata, K., Fukuyama, Y., Takayama, S., & Nakanishi, Y. (2000). A particle swarm optimization for reactive power and voltage control considering voltage security assessment. *IEEE Transactions on Power Systems*, *15*(4), 1232–1239.

Young, J. A., Smith, T. J., & Zheng, S. H. (2020). Call me BIG PAPA: An extension of Mason's information ethics framework to big data. *Journal of the Midwest Association for Information Systems*, *2*, 17–41.

Zamani, E. D., Griva, A., & Conboy, K. (2022). (2022). Using Business Analytics for SME Business Model Transformation under Pandemic Time Pressure. *Information Systems Frontiers*. Advance online publication. doi:10.100710796-022-10255-8 PMID:35250366

Zebari, D. A., Zeebaree, D. Q., Abdulazeez, A. M., Haron, H., & Hamed, H. N. A. (2020). Improved threshold based and trainable fully automated segmentation for breast cancer boundary and pectoral muscle in mammogram images. *IEEE Access: Practical Innovations, Open Solutions*, *8*, 203097–203116. doi:10.1109/ACCESS.2020.3036072

Zeiler & Fergus. (2014). Visualizing and understanding convolutional networks. In *European Conference on Computer Vision* (vol. 8689). Springer.

Zhang, H., Cheng, N., Zhang, Y., & Li, Z. (2021). Label flipping attacks against Naive Bayes on spam filtering systems. *Applied Intelligence*, *51*(7), 4503–4514. doi:10.100710489-020-02086-4

Zhang, X. D. (2020). Machine learning. In *A Matrix Algebra Approach to Artificial Intelligence* (pp. 223–440). Springer. doi:10.1007/978-981-15-2770-8_6

Zhang, Z., Guo, X., & Lin, Y. (2018). Trust management method of D2D communication based on RF fingerprint identification. *IEEE Access: Practical Innovations, Open Solutions*, *6*, 66082–66087. doi:10.1109/ACCESS.2018.2878595

Zheng, S., Janecek, A., & Tan, Y. (2013, June). Enhanced fireworks algorithm. In Evolutionary Computation (CEC), 2013 IEEE Congress on (pp. 2069-2077). IEEE. doi:10.1109/CEC.2013.6557813

Zheng, S., Janecek, A., Li, J., & Tan, Y. (2014, July). Dynamic search in fireworks algorithm. In Evolutionary Computation (CEC), 2014 IEEE Congress on (pp. 3222-3229). IEEE. doi:10.1109/CEC.2014.6900485

Zu, Z. Y., Jiang, M. D., Xu, P. P., Chen, W., Ni, Q. Q., Lu, G. M., & Zhang, L. J. (2020). Coronavirus disease 2019 (COVID-19): A perspective from China. *Radiology*, *296*(2), E15–E25. doi:10.1148/radiol.2020200490 PMID:32083985

About the Contributors

Hadj Ahmed Bouarara received a license degree in computer science, a master's diploma in computer modeling of knowledge and reasoning, Doctorate in web and knowledge engineering from Dr. Tahar Moulay University and HDR from Sidibe-labess university. Currently, Bouarara is an associate Professor of artificial intelligence and information retireval at Saida University, Algeria since October 2017. Bouarara is a Senior Associate Professor, Director of Ph.D. (since June 2020), the supervisor of several master's projects, and a member of different PRFU projects called "Bioinspired for Information Retrieval" (2014 - 2018) at GeCoDe laboratory. Bouarara is a valid instructor at Huawei ICT Academy on routing and switching and AI traks since 2021. He did some seminars in deep learning and bioinspiration in different universities such as Saida, Sidibelabess, and Telemcen. Previously he was a teacher at Elbadr Institution and also an active member of the team bio mimicry and the Saida information technology club (SITC). Bouarara is a part of the project metaheuristic for information retrieval in the GeCoDe laboratory and has more than 45 publications (books, book chapters, articles, conference papers…). He was a member in the organisation and PC member of different international conferences such as CIIA, CNTA, JERI, ICTAACS'21, ICCSA'21 and other conferences. Before becoming an academic, he have achieved 80% on average and more in 15 computer science certifications. Bouarara has a score of 10 from the Google scholar H index and 16 on Research Gate. He is an associate-Editor of IJOCI and IJIRR, an editorial advisory in JITR andIJIRR, and a reviewver in IJSIR, IJORIS, IJSSCI.

* * *

Bekaddour Abdel Madjid received a license degree in computer science specilité information system, a master's's diploma in computer modeling of knowledge and reasoning, and is a PhD student in automatic language processing from mascara university. Madjid's subjects of research are various, such as: artificial intelligence, big data, security and augmented reality.

Marion Adebiyi received a BSc. degree in computer science from the University of Ilorin, Kwara State, Nigeria in 2000. Her MSc. and Ph.D. degree are also in computer science, and bioinformatics options from Covenant University, Ota, Nigeria in 2008 and 2014 respectively. She is a senior lecturer in the Computer Science Department of Landmark University and Covenant University. She authors two books, over ten chapters, and more than 50 articles, with over 25 conferences and workshops. Dr. Adebiyi heads the H3Africa projects coordinating Division, and Entomology and Data Management Division of Covenant University Bioinformatics Research (CUBRe) group, and her research interests include bioinformatics, genomics, proteomics, and Organism's inter-pathway analysis. From 2002 to 2004, she was a web content developer/ IT Technical Officer at the World Mission Agency, Canaan land Nigeria. Since 2006, she joined the Computer and Information Sciences Department of Covenant University as a Graduate Assistant and rose through the ranks to a Senior Lecturer position in 2017, and currently a Reader. She was a Post-Doctoral Fellow at the Department of Information Technology, Durban University of Technology, Durban. South Africa. She authors about ten books and book chapters, more than 60 research articles in highly indexed journals, and has attended over 45 conferences and workshops. She has graduated over 150 Undergraduates, 8 master's's / PGDs and co-supervised 2 Doctorate degree students in computer science and Bioinformatics Her research interests include bioinformatics, genomics, proteomics, transcriptomics, and Organism inter-pathway analysis, she is involved in developing and implementing approaches and methods used in genetics research to associate specific genetic variations with diseases and traits, with interest in Infectious diseases of African populations, pathogens, hosts, and vectors. She serves as a reviewer for many peer-review journal outlets. She was the West African President of the Regional Student Group Student Council (WA-RSGSC) of the International Society for Computational Biology (ISCB) from 2008-2013, served as the Treasurer for the African Society for Computational Biology and Bioinformatics (ASBCB) from 2015-2020 and currently the Treasurer for Nigerian Bioinformatics and Genomics Network (NBGN).

Nora Alicherif received a license degree in computer science specialty information system modeling language, master's diploma in security information and cryptography (SIC), actually i am a phd student in big data security from sidi belabess university. Alicherif's subjects of research are from different domains such as: security, cryptography, internet of things, and blockchain.

Cheimaa Bentadj received a bachelor's degree in computer science and a master's degree in computer science modeling of knowledge and reasoning from the computer science department of Dr. Moulay Taher University of Saida Algeria. Her research interests include data mining, text mining, deep learning, convolutional neural networks, machine learning, and computer vision. She has published in journals such as the International Journal of Applied Evolutionary Computation (IJAEC), and the International Journal of Security and Privacy in Pervasive Computing (IJSPPC).

V. Vidya Chellam is an Assistant Professor in the department of management studies, Directorate of distance education, at Madurai Kamaraj University, Madurai.

Abeer S. Desuky received the B.Sc. degree in science, in 2003, and the M.Sc. and Ph.D. degrees in computer science, in 2008 and 2012, respectively. She is currently an Associate Professor in computer science with the Mathematics Department, Faculty of Science, Al-Azhar University, Cairo, Egypt. She has published several research papers in the field of AI, machine learning, metaheuristic optimization, and data mining and analysis. Supervisor of some master's and Ph.D. theses. She is a reviewer in many Scopus-indexed journals, such as IEEE ACCESS, Egyptian Informatics Journal, and Advances in Systems Science and Applications (ASSA).

Lamiaa El Bakrawy received the M.S. and Ph.D. degrees in computer science from the Faculty of Science, Al-Azhar University, Egypt, in 2009 and 2012, respectively. Her master's topic was optimization using swarm intelligence. Her Ph.D. topic was machine learning in image authentication. She is currently an Associate Professor with the mathematics and computer department, Faculty of science, at Al-Azhar University. She has published many articles in well-known international journals. Her research interests include image processing, social network analysis, computational intelligence, machine learning, metaheuristic optimization, and information security. She is also a program committee member in various international conferences.

K. V. B. Ganesh is a faculty of OB and HR, with 14 years of experience in academics.

Ankur Gupta has received the B.Tech and M.Tech in Computer Science and Engineering from Ganga Institute of Technology and Management, Kablana, affiliated with Maharshi Dayanand University, Rohtak in 2015 and 2017. He is an Assistant Professor in the Department of Computer Science and Engineering at Vaish College of Engineering, Rohtak, and has been working there since January 2019. He has many publications in various reputed national/ international conferences, journals, and online book chapter contributions (Indexed by Scopus, ESCI, ACM, DBLP, etc). He is doing research in the field of cloud computing, data security & machine learning. His research work in M.Tech was based on biometric security in cloud computing.

Rita Komalasari is a lecturer at YARSI University. Her current work is focused on the ethical consideration of using artificial intelligence (AI) in medicine.

Ibidun Obagbuwa is an NRF-rated researcher. She obtained a Ph.D. in computer science from the University of KwaZulu-Natal (UKZN), South Africa in 2015. She is currently a senior lecturer in the Department of Computer Science and Information Technology at Sol Plaatje University, where she is diligently performing her academic roles and serving in different capacities at the university, school, and department levels. Her research interest includes machine learning, digital image processing, computer vision, artificial intelligence, bioinformatics, and optimization. She is a diligent scholar, and her research has had a great impact on recent advances in the computer science and data science field. Her research has been cited by many computer scientist researchers across the globe with over 600 citations. Furthermore, she is participating in academic citizenship within the university and externally. She has examined both Ph.D. and master's theses locally and internationally. She has a good peer review record with different journals and professional organizations. She is a member of the South African institute of computer scientist and information Technologies (SAISCIT) professional body. She is an associate member of the National Institute for Theoretical and Computational Sciences (NITheCS) and the Centre of Excellence in Mathematical and Statistical Sciences (Coe-mass).

Sabyasachi Pramanik is a professional IEEE member. He obtained a PhD in Computer Science and Engineering from the Sri Satya Sai University of Technology and Medical Sciences, Bhopal, India. Presently, he is an Assistant Professor, Department of Computer Science and Engineering, Haldia Institute of Technology, India. He has many publications in various reputed international conferences, journals, and online book chapter contributions (Indexed by SCIE, Scopus, ESCI, etc). He is doing research in the field of artificial intelligence, data privacy, cybersecurity, network security, and machine learning. He is also serving as the editorial board member of many international journals. He is a reviewer of journal articles from IEEE, Springer, Elsevier, Inderscience, IET, and IGI Global. He has reviewed many conference papers, has been a keynote speaker, session chair and has been a technical program committee member in many international conferences. He has authored a book on Wireless Sensor Network. Currently, he is editing 6 books from IGI Global, CRC Press EAI/Springer and Scrivener-Wiley Publications.

Praveenkumar S. is an Assistant Professor at the Centre for Tourism and Hotel Management, at Madurai Kamaraj University, Madurai, Tamilnadu, India.

Index

E

F

G

H

I

K

M

O

P

S

T

U

V

W

www.ingramcontent.com/pod-product-compliance
Lightning Source LLC
LaVergne TN
LVHW080859300926
849162LV00003B/487